SpringerBriefs in Applied Sciences and Technology

More information about this series at http://www.springer.com/series/8884

Arief Suriadi Budiman

Probing Crystal Plasticity at the Nanoscales

Synchrotron X-ray Microdiffraction

 Springer

Arief Suriadi Budiman
Singapore University of Technology
    and Design
Singapore
Singapore

ISSN 2191-530X   ISSN 2191-5318   (electronic)
SpringerBriefs in Applied Sciences and Technology
ISBN 978-981-287-334-7   ISBN 978-981-287-335-4   (eBook)
DOI 10.1007/978-981-287-335-4

Library of Congress Control Number: 2014957300

Springer Singapore Heidelberg New York Dordrecht London

Printed on acid-free paper

Springer Science+Business Media Singapore Pte Ltd. is part of Springer Science+Business Media
(www.springer.com)

Acknowledgments

When I finally decided in April 2002 to become a mid-career student in a Ph.D. program at Stanford University—from which this book is originated—little did I know the insurmountable emotional "mountain" that I was about to have to move. But the Good Book was right. "If you just have faith as small as a mustard seed, nothing will be impossible" (*Matthew 17:20*). If I am standing where I am right now, that is only by that faith—no matter how very small it might have felt at times during these past five years—and with the help and sacrifice of many people around me.

I would first and foremost like to thank my former Ph.D. advisor at Stanford, Professor William D. Nix. I remember speaking to him on the phone for the very first time in Spring 2001 to discuss financial support to enable me to come to Stanford. Even in that first conversation, I was already struck by his warmth, passion and kind encouragements, even though he barely knew me at the time. I realize now how calling his office that afternoon was indeed a stroke of luck in my lifetime. He has certainly been my greatest advisor; the closest I have ever been and will probably ever be to such an inspiring man of science of such world stature, my academic father; but more importantly he has also been like my own father, supporting me in literally every aspects of my life. He is a truly fine man that I am so privileged to get to know and work with for the past five years.

In the end of my penultimate year at Stanford, I lost another research advisor of mine, Professor Jamshed R. Patel, to whom I am very much indebted for his guidance, inspiration and enthusiasm. I could not have come to Stanford and eventually realized my dream, if it were not for his hands in the beginning. I will always remember his gentle style of teaching, quiet, and unassuming with the keenest of intellects, but always with the time to help and encourage me.

The work that led to the publication of this book would have been nowhere as intellectually intriguing without the guidance of another person of such rare talent and great enthusiasm for doing experimental research. I owe it to Dr. Nobumichi Tamura of the Advanced Light Source (ALS), Berkeley Lab, for literally teaching me everything that I know about synchrotron X-ray microdiffraction and leading me in an exploration deep into the wonderful world of the reciprocal space. It has certainly been a great pleasure and a magical learning experience working with him.

Parts of this book have been close collaborations with colleagues in industry and other university, whom I truly admire. In particular, Dr. Paul R. Besser, Dr. Christine Hau-Riege, and Dr. Amit Marathe of Advanced Micro Devices, Inc. (AMD), and Professor Young-Chang Joo of the Seoul National University (SNU) have instilled in me the confidence that I need to succeed as a research professional. It is, I believe, with their close guidance and support that I achieved what I have achieved in 2006. Two invited talks at the Materials Research Society (MRS) meetings, and one graduate student award. Dr. Jose Maiz and Dr. Kaustubh Gadre of Intel Corporation are also acknowledged for their great support and engaging interactions.

I would also like to acknowledge the Singapore University of Technology and Design (SUTD) where I am currently affiliated as a Tenure-Track Faculty Member at the Engineering Products Design (EPD) Pillar. The research group that I am establishing here—the Xtreme Materials Lab by Design (xml.sutd.edu.sg) is something that I am very proud of and consists of promising young men and women of materials science, technology and design. In particular, I would like to thank one of my Ph.D. students—Ihor Radchenko—who has helped me tremendously in the final edit/revision/update and production of the manuscript of this book. After all, his careful hands and eyes are what make this book a book (from a Ph.D. dissertation at Stanford University)!

I would also like to extend my most sincere gratitude to my family in Indonesia, my Mom and my brother and sisters, and their spouses, for their constant encouragements and support—without which this book would have been close to impossible. Their unconditional love and devoted support for me is a constant source of strength. My late father remains an inspiration to me, constantly fueling the burning desire in my chest to make sure that our next generations could have a better, more meaningful life. My nephews and nieces have been an additional source of joy. I must also thank my in-laws in Indonesia. In particular, my mother-in-law, who, during this time, has visited us a few times to provide companionship and support for me and my wife.

Finally, words fail me when I have to express my utmost gratitude and heartfelt feeling of indebtedness to my wife, Grace Tanja. The sacrifice that she has made has enabled me to realize my dream, and any hardship and difficulty that I faced in my research pales in comparison to hers in making us a home—a home that has also warmly welcomed the arrivals of our two beautiful boys during this journey at Stanford. This book—the fruit of my labor, of my heart and of every inch of my dream, aspiration and passion—I earnestly dedicate for her and for the sacrifice that she has so selflessly made for us, her family. Our first son, Ethan, has been a source of pleasure and pride—the apple of our eyes. He has also been my most devoted admirer—to him, I am a superhero—and a constant source of joy and laughter, which after the rigor of graduate school days, I could certainly use a dose of those. Having my wife, Grace, my first son, Ethan and our newest addition to our family, Alexander, is indeed the Lord's blessing in my life.

Singapore, October 2014   Arief Suriadi Budiman

Contents

Chapter 1
Introduction

Abstract Small scale plasticity plays an important role in the modern electronics. The µSXRD technique offers the unique capability to study the plastic evolution of the grains in the interconnect lines during electromigration (in situ) at the submicron resolution. These experiments provide useful insights and may also provide important practical implications, as will be discussed in greater detail, for the fundamental understanding of the electromigration degradation mechanisms, as well as for the industry critical assessment methodologies of electromigration device lifetime. The technique can also be used to provide the key tool to probe the plastic behavior of the materials at small scales under the mechanical load. Understanding and controlling plasticity and the mechanical properties of materials on this scale could thus lead to new and more robust nanomechanical structures and devices.

Keywords µXRD · Small scale · Plasticity · Synchrotron · Electromigration · Cu interconnect · Size effect · Stress gradient · Strain gradient

1.1 Small Scale Plasticity

The development of the modern integrated circuit and related device structures has brought about the need to understand material behavior, including mechanical properties, at the submicron and nanometer scale. In addition, nanomechanical devices will play an ever more important role in future technologies. Already transistors and interconnects at the submicron and nanometer scales are common in today's memory and microprocessors. New devices based on micro-/nano-electromechanical systems (M/NEMS) and nanotechnology are increasingly becoming a reality in the marketplace. The creation of such small components requires a thorough understanding of the mechanical properties of materials at these small length scales. Here we propose to directly examine some of the effects that arise when crystalline materials are plastically deformed in small volumes.

© The Author(s) 2015
A.S. Budiman, *Probing Crystal Plasticity at the Nanoscales*,
SpringerBriefs in Applied Sciences and Technology,
DOI 10.1007/978-981-287-335-4_1

One type of small scale structure is represented by the metal wires used in microelectronic chips as interconnects. Metal thin films patterned into micron–scale conductor lines comprise the communication network of all integrated circuits. When the electrical current density running through these increasingly smaller and smaller wires becomes large enough (in the order of MA/cm^2), atoms start to migrate, inducing voids and hillocks to form under certain circumstances and eventually resulting in the final catastrophic failure of the device. Prior to the present research, Valek et al. [1] discovered a very unusual mode of plastic deformation occurring at an early stage of electromigration in Al interconnects. The deformation geometry introduces dislocation lines predominantly in the direction of electron flow, and thus may provide additional easy paths for the transport of point defects. Since these findings occur long before any observable voids or hillocks are formed, they may have a direct bearing on the final catastrophic events of failure of the device.

In that previous study, the unique and powerful capability of synchrotron X-ray microdiffraction became clear. Utilizing submicron-focused polychromatic synchrotron beam developed in the Beamline 12.3.2 at the Advanced Light Source (ALS), Berkeley Lab, the technique proved advantageous as a local probe of mechanical behavior and plastic deformation in small scale devices. Furthermore, with this facility, in situ electromigration experiment can now be done, which is a rare opportunity that is not always practical with other characterization techniques. This capability enables us to investigate the evolution of the structure of the crystals as they deform due to the enormous wind force of electrons moving from one end of the interconnect line to the other. This is an important piece of information for the fundamental understanding of electromigration degradation processes in interconnects.

The mechanical behavior of materials in small dimensions (submicron and nanoscale) often deviates considerably from the behavior of bulk materials. In the macroscale (bulk), the mechanical properties of materials are commonly described by single valued parameters (e.g. yield stress, hardness, etc.), which are largely independent of the size of the specimen. However, as specimens are reduced in size to the scale of the microstructure, their mechanical properties deviate from those of bulk materials. For example, in thin films—where only one dimension, the thickness, reaches the micron scale and below—the flow stress is found to be higher than its bulk value, and becomes even higher as the film gets thinner. These size effects are usually attributed to grain size hardening [2–6] and to the confinement of dislocations within the film by the presence of the substrate and in some cases the passivation [7–13].

In nanoindentation, strength has been known to depend inversely with indentation depth—the smaller the depth of indentation, the harder the crystal. This effect has been attributed to strain gradients or geometrically necessary dislocations (GNDs) [14–28]. In other small scale structures, like submicron single crystal pillars subjected to uniaxial compression, however, this size effect has been observed even though the deformation imposes no strain gradients [29–31]. These different kinds of size effects observed as the scale of deformation tends toward the submicron and nanoscales, thus suggest that plasticity is no longer governed by the classical plasticity model.

In the first part of this book, the unique capability of the synchrotron white-beam X-ray microdiffraction is discussed, not only to study the mechanical behavior of interconnect lines in situ and at submicron resolution, but also as a local probe to detect and measure the densities of geometrically necessary dislocations (GNDs). It is therefore, a very suitable technique, to study the role of geometrically necessary dislocations and strain gradients in small scale plasticity, which could contribute to the understanding of the size effects in crystal hardening mechanisms.

As we have seen, synchrotron white beam X-ray microdiffraction provides a new information that would be difficult to obtain with other characterization techniques. Most importantly, it provides a direct way to measure strain gradients quantitatively. Hence, in the following section, we introduce the technique itself and explain how it has been useful in both the electromigration study, as well as the size effect investigation.

1.2 White-Beam X-ray Microdiffraction as Plasticity Probe

Synchrotron white-beam X-ray microdiffraction is essentially an X-ray Laue diffraction technique. Its unique feature stems from the fact that the X-ray beam comes from a synchrotron source, which is orders of magnitude brighter than the laboratory X-ray source, and thus can be focused into a submicron spot size. This capability enables characterization of materials and their mechanical properties at high (submicron) spatial resolution. The polychromatic characteristic of the synchrotron radiation makes it sensitive to local lattice curvature or rotation in the crystals under consideration.

Since strain gradients, or equivalently, the geometrically necessary dislocations are directly related to the local lattice curvature, this technique has been suitable for probing plasticity at small scales. This sensitivity to local lattice curvature is related to the continuous range of wavelengths in a white X-ray beam, allowing Bragg's Law to be satisfied even when the lattice is locally rotated or bent, resulting in the observation of streaked Laue spots.

X-ray diffraction is a powerful, century-old technique routinely used with laboratory and synchrotron sources to study the structural properties of materials. Compared to electron probes, X-rays offer the advantages of deeper penetration depths (so that bulk and buried samples can be investigated), and of virtually no sample preparation and measurement under a variety of different conditions (in air, liquid, gas, vacuum, at different temperatures and pressures). The technique of synchrotron-based white beam X-ray diffraction is one of the few methods for detecting and measuring the densities of GNDs in crystalline materials after deformation without the need to destroy the sample. Electron microscopy techniques, such as TEM, can also be used to detect presence of dislocations in small volumes, but the thinning required to obtain electron transparency could alter the defect structure being observed.

The synchrotron technique of scanning white beam X-ray microdiffraction has been described in a complete manner elsewhere [32], and will be reviewed in the Chap. 2 of this book. The use of this technique involves scanning the sample with the focused X-ray beam at submicron resolution, thus gaining structural information about the crystal and its defects in the diffracted volume through the shapes of the Laue diffraction peaks. Using this approach, we can monitor the change in the Laue diffraction peaks before and after deformation, sometimes even during the deformation (in situ). A quantitative analysis of the Laue peak widths then allows us to estimate the density of GNDs in the sample. The absolute number of geometrically necessary dislocations in the crystal can then be determined using the relevant dimensions of the sample. A comparison of the numbers of geometrically necessary dislocations before and after, or even during, the deformation provides information about the change in microstructure associated with plastic deformation.

Having described the comparative advantage of this technique to map mechanical properties of interconnects under deformation, we next introduce the electromigration phenomenon.

1.3 Electromigration in Metallic Interconnects

We have seen from the last section that the sensitivity to local lattice rotation or curvature is the critical feature of the white-beam X-ray microdiffraction technique. This has proven to be useful in the study of the early stages of electromigration failure in interconnect lines, wherein lattice bending and GNDs are created by electromigration processes [33, 34], as will be discussed in Chaps. 3–5 in this book. This becomes especially important because this early plastic behavior, exhibited by both Al and Cu polycrystalline interconnect lines, may have a direct bearing on the final failure stages of electromigration.

1.3.1 Electromigration Fundamentals

Electromigration (EM) is a phenomenon that occurs when extremely high current densities ($j \sim 10^6$ A/cm^2) lead to mass transport within integrated-circuits [35]. Failures of the interconnects can be caused by open circuit voiding or short circuits caused by extrusion of metal from the line. Mechanical considerations have influenced the understanding of EM ever since the discovery that the directed diffusion of metal atoms, due to the momentum transfer from the electrons, can lead to stresses in the conductor line [36, 37].

The most well accepted mechanism is one where the EM drift is assumed to be concentrated along the grain boundaries running parallel to the line. If the flux continuity is locally disturbed, in the most extreme case by blocking boundaries (across the conductor line), then the accumulation of atoms (removal of vacancies)

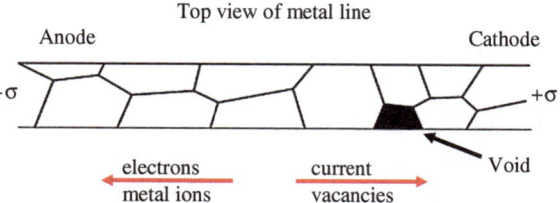

Fig. 1.1 Electromigration in metallic interconnects

at the anode end will set up a hydrostatic compressive stress; likewise, a hydrostatic tensile stress will develop at the cathode end because of the removal of atoms (accumulation of vacancies) there, as shown in Fig. 1.1. Such gradients have been measured experimentally [37] and magnitudes of several hundred MPa (up to 1 GPa) are considered typical. These "electromigration stresses" superimpose on any thermal stresses that may be present in the line. These stresses can have damaging effects on the line. Tensile stresses can initiate nucleation and drive growth of voids in the line. Compressive stresses may lead to extrusion of metal in the line and cracking of the passivation.

Although passivation may crack and thus be the source of failure of interconnect lines, passivated metal lines have longer lifetime compared to unpassivated ones. Blech [36–39] discovered that the gradients in the hydrostatic stress along the line influence the electromigration kinetics. They cause gradients in the chemical potential for atoms (or vacancies), which in turn drive a diffusional flux that opposes the drift due to EM. He termed the stress-induced driving force the "back flux" because it acts to oppose the electron wind force. When this effect is incorporated in the overall kinetics, we have the following expression for the electromigration flux

$$J_{EM} = \frac{Dc}{kT}\left(eZ^*\rho j - \Omega\frac{\Delta\sigma}{L}\right), \qquad (1.1)$$

where D is atomic diffusivity, c is the concentration of atoms, k is Boltzmann's constant, T is temperature, Z^* is the effective charge number, e is the fundamental electron charge, ρ is resistivity, j is current density, Ω is the atomic volume, and $(\Delta\sigma/L)$ is the local gradient in the hydrostatic stress along the length of the line/segment L.

As can be seen clearly from Eq. 1.1, there might be circumstances where the two terms on right hand side of the equation cancel each other out, and create a steady state ($J_{EM} = 0$). In particular, when interconnects are below a critical length, known as the Blech length, L_c, the stress-induced flux can actually cancel out the electron wind flux. Thus, interconnects below the Blech length will not fail, and known as the "immortal line." Similarly, for a given interconnect length L, the resistance change due to EM damage will cease below a certain critical current density, j_c.

1.3.2 Electromigration Degradation Mechanisms in Cu
Interconnects

While the electromigration phenomenon in Al interconnects has been widely studied [40–42], the general mechanisms of electromigration in Cu interconnects has not been as thoroughly examined. There has been an increased focus on replacing Al-based interconnects with Cu interconnects in industry, due to the scaling-related more aggressive requirements for an interconnect material with much lower resistivity. The use of copper combined with a low-k dielectric material is expected to significantly reduce the RC delay of integrated circuits [43], especially the component associated with narrow interconnect lines.

Because Cu has a higher melting temperature and therefore lower atomic diffusivity than Al, it is expected to have a substantially improved resistance to electromigration and electromigration-induced failure [44]. However, observed reliability improvements for Cu have been generally less than expected. It has been suggested that in Cu [45], unlike Al, interfacial self-diffusion is generally faster than grain-boundary self-diffusion, so that interfaces, especially the top interface between Cu and its top capping materials [45], provide high diffusivity paths that short-circuit the usual grain boundary paths. Current Cu technology involves the use of refractory metal liner layers below and on the sides of the Cu line (usually TaN), and a layer of SiN on the top of the Cu lines (capping layer) to improve adhesion between Cu and dielectric. The reliability of Cu lines has been known to improve with increased interface adhesion, especially the top interface with the capping layer. This is because increased interface adhesion impedes interfacial self-diffusion, and thus halts the fast diffusivity paths in electromigration in Cu.

It has been well-established that grain boundaries (when they are available) provide the fastest diffusion path in Al interconnects [41]. Therefore, microstructure plays a key role in the electromigration lifetime of Al interconnects, with interconnects that have bamboo microstructures outliving their polygranular counterparts, which have grain sizes equal to or less than the linewidth, by orders of magnitude [42]. That is, lines with bamboo structures in which the grain boundaries are oriented perpendicular to the direction of electron flow, have lower effective diffusivities relative to polygranular structures, in which there are grain boundaries that are oriented with boundaries lying in the direction of electron flow. Fully bamboo grain structures can result from grain growth induced by postpatterning annealing and do not result from patterning large-grained films or from recrystallization of patterned films.

The role of microstructure in the electromigration of Cu interconnects is less clear than the role of microstructure in Al. It has even been proposed that surface diffusion in Cu dominates over grain boundary diffusion [46] even in interconnects with polygranular structures. It is also known that the development of a stress gradient in the interconnect, which necessarily attends electromigration, can impede or completely halt the process [37]. While plastic deformation is expected to occur during electromigration when the stress gets high enough, quantitative information on plasticity during in situ electromigration tests is difficult to obtain, because of the size of the structures and of the local nature of the phenomenon.

A great deal of research has been conducted in an attempt to understand the role of stress and stress gradients during EM and several models have been proposed [47–49]. Experimental verification of these models has proven difficult due to the challenge of measuring stress in passivated interconnect structures with the necessary spatial resolution. With the X-ray microdiffraction technique described in the last section, local probing capability with high spatial resolution can be achieved and local stress/strain and plastic deformation measurement at the polycrystals level is possible.

1.4 Size Effects in Crystalline Materials

Plastic deformation in small volumes requires higher stresses than are needed for plastic flow of bulk materials. There are various effects, both extrinsic and intrinsic, that seem to be responsible for this observation. The size effects observed in thin metal films arise from the constraints of surrounding layer or from the micro-structural characteristics of the thin films. These kinds of size effects have been called the extrinsic effects [14]. The intrinsic effects, in contrast, are those size effects that arise in small, unconstrained single crystals under deformation. There are two possible sources of size effects that have been identified: strain gradients and dislocation starvation.

In nanoindentation experiments, where the length-scale of the deformation reaches the microstructural length-scale of the material, the governing relations between stress and strain deviate from the classical laws that apply to bulk mate-rials. At small depths of indentation the hardness of crystalline materials is usually higher than that of large indentations, as illustrated in Fig. 1.2a. This indentation size effect is an intrinsic one, and has been explained using the concept of geo-metrically necessary dislocations (GNDs) and strain gradients [15–28].

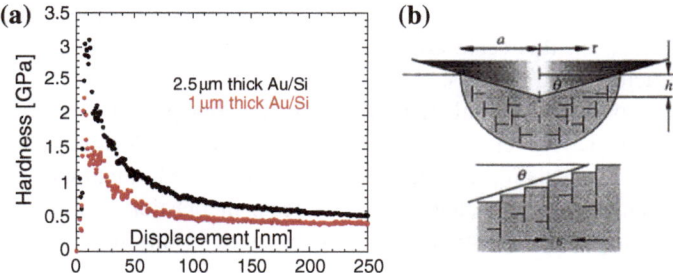

Fig. 1.2 Indentation size effects (ISE): **a** hardness versus displacement data for two thicknesses of Au thin films on Si substrate (courtesy of Lilleodden [54]); **b** geometrically necessary dislocations created by a rigid indentation

According to this picture, the hardness increases with decreasing depth of indentation because the total length of geometrically necessary dislocations forced into the solid by the self-similar indenter scales with the square of the indentation depth, while the volume in which these dislocations are found scales with the cube of the indentation depth (Fig. 1.2b). This leads to a geometrically necessary dislocation density that depends inversely on the depth of indentation. The higher dislocation densities expected at smaller depths leads naturally to higher strengths through the Taylor relation [50], and this leads to the indentation size effects (ISE).

A different kind of intrinsic size effect is observed when single crystalline materials are deformed homogenously, without strain gradients. Uchic et al. [29] and others [30, 31] have shown that micro pillars of various metals with diameters in the micron range, subjected to uniaxial compression, are much stronger than bulk materials. For example, micro pillars of gold ranging in diameter between 200 nm and several microns have been found to have compressive flow strengths as high as 800 MPa, a value ~ 50 times higher than the strength of bulk gold [30, 31]. The accounts of strain gradient plasticity, as discussed above, appear to break down for the case of micro pillar compression because the deformation is essentially uniform.

Greer and Nix [31] suggested that the high strengths of sub-micron pillars of gold might be controlled by a dislocation starvation hardening process. In this mechanism and for small enough pillars, the mobile dislocations have a higher probability of annihilating at a nearby free surface than of multiplying and being pinned by other dislocations. When the starvation conditions are met, plasticity is accommodated by the nucleation and motion of new dislocations rather than by motion and interactions of existing dislocations, as in the case of bulk crystals.

With the increasing need from industry to develop materials of high mechanical performance at small length scales, a good understanding of such material response under deformation has become important since many of these responses, as we have seen, are dependent both extrinsically as well as intrinsically on the behavior of structural entities at this scale (grain boundaries, inclusions, intrinsic intra- and inter-granular stress distribution). There is limited amount of experimental data at these scales due to the lack of suitable techniques. This has long prevented modeling material behavior at these length scales, which in turn has prevented progress in developing a systematic link of materials properties from the macroscopic to the microscopic. Understanding and controlling plasticity and the mechanical properties of materials on this scale could thus lead to new and more robust nanomechanical structures and devices.

In order to provide a useful framework for the work presented in this book, the principles of the Taylor relation between flow stress and dislocation density [29] is first presented; this is then followed by a discussion of the limitation of this classical flow-stress relation. One account of size effects in plasticity—the Nix and Gao model of strain gradient plasticity [19]—is reviewed in this section.

1.4.1 Classical Flow-Stress Relationship: The Taylor Relation

The Taylor relation states that the flow stress of a material is proportional to the square root of the density of dislocations. By recognizing that the shear stress resulting from the self-stress field of a dislocation varies as the inverse of the distance from the dislocation, and that the average spacing of dislocations is defined by the reciprocal of the square root of the dislocation density, ρ, Taylor argued that the stress required to move a dislocation past another dislocation, τ, is given by:

$$\tau = \frac{\mu b}{2\pi} \sqrt{\rho}, \tag{1.2}$$

where μ is the shear modulus of the material, b is the Burgers vector.

The more general form of this relation for the case of a dislocation moving in a forest of dislocations is given by:

$$\tau = \alpha \mu b \sqrt{\rho}, \tag{1.3}$$

where α is the Taylor coefficient, which depends on the character of the dislocation forest [50–53]. The important implication of this relation is that the flow stress is always enhanced by the presence of dislocations.

However, this statement cannot always be true. As the scale of deformation tends toward the scale of microstructural features (the spacing of dislocations, grain sizes) this classical relation may no longer hold. This has been illustrated clearly by Lilleodden using Fig. 1.3 (courtesy of Lilleodden [54]). Figure 1.3a illustrates that when the volume is large relative to the spacing of dislocations, the Taylor relation

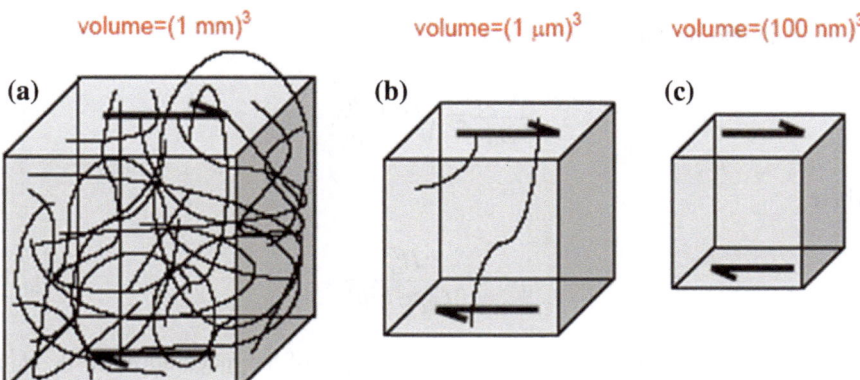

Fig. 1.3 Illustration of the limitation of Taylor relation: **a** deformed volume $\gg \rho^{-3/2}$, **b** deformed volume $\sim \rho^{-3/2}$, **c** deformed volume $< \rho^{-3/2}$; in each case, the same dislocation density is assumed; (courtesy of Lilleodden [54])

is expected to fully describe the flow stress. In Fig. 1.3b, the volume is close to the spacing of dislocations, such that the averaging implied by Eq. 1.3 is no longer applicable, and consequently a deviation from the Taylor relation is expected. Finally, in the limit of a deformed volume that is less than the cube of the dislocation spacing, as in Fig. 1.3c, plasticity may be controlled by the creation of dislocations, which requires near-theoretical stresses. Thus, the Taylor relation would obviously not be applicable in this case. The flow stress in the case described by Fig. 1.3c should be very high. This is one limitation of the classical flow-stress relationship described by Taylor relation. Clearly, the length scale of the deformation needs to be considered when describing the flow stress.

1.4.2 The Nix and Gao Model of Strain Gradient Plasticity

Indentation size effects (ISE) have been widely observed. Crystalline materials display a strong size dependence when they are indented over a scale from a fraction of a micron to tens of microns: the smaller the indentation depth, the higher the hardness, such as shown in Fig. 1.2a. Such behavior cannot be explained using classical plasticity as we have discussed in the last sub-section (Taylor relation). Following the work of Stelmashenko et al. [15] and De Guzman et al. [16], Nix and Gao [19] provided a simple explanation for this depth-dependent hardness, in terms of the geometrically necessary dislocation density as a function of indentation depth. Durst et al. [27] as well as Feng [55] later modified the model to account primarily for the fact that the plastic zone radius is not equal to the contact radius, as Nix and Gao had assumed. Still the revised model takes the form:

$$\frac{H(h)}{H_0} = \sqrt{1 + \frac{h_0}{h}}, \tag{1.4}$$

which can also be shown equivalently:

$$\frac{H(h)}{H_0} = \sqrt{1 + \frac{\rho_G}{\rho_S}}, \tag{1.5}$$

where

$$\rho_S = \frac{H_0^2}{3C_H^2 \alpha_t^2 \mu^2 b^2}. \tag{1.6}$$

$H(h)$ is the hardness as a function of h, the depth of the indentation, while H_0 is the limit of the hardness when the indentation depth (h) becomes indefinitely large, and h_0 is a material length scale. In Eq. 1.6, C_H is a constant associated with the plastic zone size, α_t is the Taylor constant, μ is shear strength and b is the magnitude of Burgers vector, ρ_S, ρ_G are the densities of statistically stored and geometrically necessary dislocations accordingly.

Equation 1.5, thus, implies depth-dependent ρ_G (as ρ_S is nominally constant), or in other words depth-dependent strain gradients. The correlation between the hardness numbers and the associated ρ_G had been derived in a complete manner elsewhere [55], and with some rearrangements, the final form such as shown in Eq. 1.7 can be used;

$$\rho_G = \frac{H_0^2}{3C_H^2 \alpha_t^2 \mu^2 b^2}\left\{\left(\frac{H}{H_0}\right)^2 - 1\right\}. \tag{1.7}$$

Mechanism-based strain gradient (MSG) plasticity is a model of hardening in which the Taylor relation is adopted as a founding principle:

$$\tau = \alpha \mu b \sqrt{\rho_{total}(h)} = \alpha \mu b \sqrt{\rho_G(h) + \rho_S} \tag{1.8}$$

The total dislocation density, ρ_{total}, is the sum of the statistically stored dislocations (SSDs), ρ_S, and the geometrically necessary dislocations (GNDs), ρ_G, which is a function of h. This phenomenological model modifies the classical Taylor relation by adding the strain gradient term which depends on deformation length scales.

References

1. Valek BC, Bravman JC, Tamura N et al (2002) Electromigration-induced plastic deformation in passivated metal lines. Appl Phys Lett 81:4168–4170
2. Arzt E (1998) Size effects in materials due to microstructural and dimensional constraints: a comparative review. Acta Mater 46:5611–5626
3. Keller RM, Baker SP, Arzt E (1998) Quantitative analysis of strengthening mechanisms in thin Cu films: effects of film thickness, grain size, and passivation. J Mater Res 13:1307–1317
4. Yu YW, Spaepen F (2003) The yield strength of thin copper films on Kapton. J Appl Phys 95:2991–2997
5. Kraft O, Hommel M (2001) Deformation behavior of thin copper films on deformable substrates. Acta Mater 49:3935–3947
6. Huang H, Spaepen F (2000) Tensile testing of free-standing Cu, Ag and Al thin films and Ag/ Cu multilayers. Acta Mater 48:3261–3269
7. Nix WD (1989) Mechanical properties of thin films. Metall Trans A 20:2217–2245
8. Nix WD (1998) Yielding and strain hardening of thin metal films on substrates. Sci Mater 39:545–554
9. von Blanckenhagen B, Gumbsch P, Arzt E (2003) Dislocation sources and the flow stress of polycrystalline thin metal films. Phil Mag Lett 83:1–8
10. Arzt E, Dehm G, Gumbsch P et al (2001) Interface controlled plasticity in metals: dispersion hardening and thin film deformation. Prog Mater Sci 46:283–307
11. Nix WD, Leung OS (2001) Thin films: plasticity. In: Buschow KHJ et al (eds) Encyclopedia of materials: science and technology. Elsevier, Oxford, p 9262

12. Han SM, Philips MA, Nix WD (2009) Study of strain softening behavior of Al-Al$_3$ S$_c$ multilayers using microcompression testing. Acta Mat 57:4473-4490
13. Pant P, Schwartz KW, Baker SP (2003) Dislocation interactions in thin FCC metal films. Acta Mater 51:3243–3258
14. Nix WD, Greer JR, Feng G (2007) Deformation at the nanometer and micrometer length scales: effects of strain gradients and dislocation starvation. Thin Solid Films 515:3152–3157
15. Stelmashenko NA, Walls MG, Brown LM et al (1993) Microindentations on W and Mo oriented single crystals: an STM study. Acta Metall Mater 41:2855–2865
16. De Guzman MS, Neubauer G, Flinn P et al (1993) The role of indentation depth on the measured hardness of materials. Mater Res Soc Proc 308:613
17. Ma Q, Clarke DR (1995) Size dependent hardness of silver single crystals. J Mat Res 10:853–863
18. Poole WJ, Ashby MF, Fleck NA (1996) Micro-hardness of annealed and work-hardened copper polycrystals. Scripta Mat 34:559–564
19. Nix WD, Gao H (1998) Micro-hardness of annealed and work-hardened copper polycrystals. J Mech Phys Solids 46:411–425
20. Gao H, Huang Y, Nix WD (1999) Modeling plasticity at the micrometer scale. Naturwissenschaftler 86:507
21. Gao H, Huang WD, Nix JW et al (1999) Mechanism-based strain gradient plasticity—I. Theory. J Mech Phys Solids 47:1239–1263
22. Huang Y, Chen JY, Guo TF et al (1999) Analytic and numerical studies on mode I and mode II fracture in elastic-plastic materials with strain gradient effects. Int J Fract 100:1–27
23. Huang Y, Gao H, Nix WD et al (2000) Mechanism-based strain gradient plasticity—II. Analysis. J Mech Phys Solids 48:99–128
24. Huang Y, Xue Z, Gao H et al (2000) A study of microindentation hardness tests by mechanism-based strain gradient plasticity. J Mater Res 15:1786–1796
25. Tymiak NI, Kramer DE, Bahr DF et al (2001) Plastic strain and strain gradients at very small indentation depths. Acta Mater 49:1021–1034
26. Swadener JG, George EP, Pharr GM (2002) The correlation of the indentation size effect measured with indenters of various shapes. J Mech Phys Solids 50:681–694
27. Durst K, Backes B, Goken M (2005) Indentation size effect in metallic materials: correcting for the size of the plastic zone. Scripta Mat 52:1093–1097
28. Durst K, Backes B, Franke O et al (2006) Indentation size effect in metallic materials: Modeling strength from pop-into macroscopic hardness using geometrically necessary dislocations. Acta Mat 54:2547–2555
29. Uchic MD, Dimiduk DM, Florando JN et al (2004) Sample dimensions influence strength and crystal plasticity. Science 305:986–989
30. Greer JR, Oliver WC, Nix WD (2005) Size dependence of mechanical properties of gold at the micron scale in the absence of strain gradients. Acta Mater 53:1821–1830
31. Greer JR, Nix WD (2006) Nanoscale gold pillars strengthened through dislocation starvation. Phys Rev B 73:245410
32. Tamura N, MacDowell AA, Spolenak BC et al (2003) Scanning X-ray microdiffraction with submicrometer white beam for strain/stress and orientation mapping in thin films. J Synchrotron Rad 10:137–143
33. Valek BC (2003) X-ray microdiffraction studies of mechanical behavior and electromigration in thin film structures. Dissertation, Stanford University
34. Budiman AS, Tamura N, Valek BC et al (2006) Crystal plasticity in Cu damascene interconnect lines undergoing electromigration as revealed by synchrotron X-ray microdiffraction. Appl Phys Lett 88:233515
35. Lloyd JF (1999) Electromigration in integrated circuit conductors. J Phys D 32:R109–R118
36. Blech IA (1976) Electromigration in thin aluminum films on titanium nitride. J Appl Phys 47:1203–1208

37. Blech IA, Herring C (1976) Stress generation by electromigration. Appl Phys Lett 29:131–133
38. Black JR (1969) Electromigration—a brief survey and some recent results. IEEE Trans Electr 16:338–347
39. Blech IA (1998) Diffusional back flows during electromigration. Acta Mater 46:3717–3723
40. Hu CK, Reynolds S (1997) CVD Cu interconnections and electromigration. Electrochem Soc Proc 97:1514
41. Joo YC, Thompson CV (1994) Analytic model for the grain structures of near-bamboo interconnects. J Appl Phys 76:7339–7346
42. Thompson CV, Lloyd JR (1993) Electromigration and IC interconnects. Mater Res Bull 18:19–25
43. Hu CK, Small MB, Kaufman F et al (1990) Copper-polyimide wiring technology for VLSI circuits. Mat Res Soc Symp Proc VLSI 369–373
44. Brown AM, Ashby MF (1980) Correlations for diffusion constants. Acta Metall 28:1085–1101
45. Hu CK, Lee KY, Gignac L et al (1997) Electromigration in 0.25 μm wide Cu line on W. Thin Solid Films 308–309:443–447
46. Hu CK, Rosenberg R, Lee KY (1999) Electromigration path in Cu thin-film lines. Appl Phys Lett 74:2945–2947
47. Korhonen MA, Borgesen P, Tu KN et al (1993) Stress evolution due to electromigration in confined metal lines. J Appl Phys 73:3790–3799
48. Gleixner RJ, Nix WD (1998) Effect of "bamboo" grain boundaries on the maximum electromigration-induced stress in microelectronic interconnect lines. J Appl Phys 83:3595–3599
49. Surh MP (1999) Threshold stress behavior in thin film electromigration. J Appl Phys 85:8145–8154
50. Basinski SJ, Basinski ZS (1979) Plastic deformation and work hardening. In: Nabarro FRN (ed) Dislocations of solids, vol 4: Dislocations in metallurgy. North-Holland Publishing Company, Oxford, p 261
51. Nabarro FRN, Basinski ZS, Holt DB (1964) The plasticity of pure single crystals. Adv Phys 13:193–323
52. Basinski ZS, Basinski SJ (1964) Dislocation distributions in deformed copper single crystals. Phil Mag 9:51–80
53. Basinski ZS (1974) Forest hardening in face centred cubic metals. Scripta Metall 8:1301–1308
54. Lilleodden ET (2001) Indentation-induced plasticity of thin metal films. Dissertation, Stanford University
55. Feng G (2005) Dissertation, Stanford University

Chapter 2
Synchrotron White-Beam X-ray Microdiffraction at the Advanced Light Source, Berkeley Lab

Abstract The µSXRD technique is developed at the Advanced Light Source, Berkeley Lab. The technique can provide a local information about the crystal structure of materials in a microscale. This became possible due to the focusing of X-rays into a micron size spot. The numerical analysis of the X-ray diffraction allows to find and quantify not only the crystal structure and lattice parameters, but also the density of geometrically necessary dislocation density, crystal bending, polygonization and rotation, and local stress/strain probing at different conditions. The physical components of the experimental setup and the experimental procedure are also described in details in this chapter.

Keywords ALS · µXRD · Beamline · Microdiffraction · White-beam · XMAS · Local plasticity · Crystal poligonization · Crystal rotation · Crystal parameter

2.1 Introduction

Traditional X-ray diffraction is a technique that has been used for almost a century for elucidating the structure of materials on the macroscopic scales (0.1–10 mm). As modern electronics, photonics and even biological devices are increasingly made in smaller and smaller scales (submicron and nanometer scales), a thorough understanding of the materials structure-properties-performance relationship at such length scales (0.1–10 µm) has become critical, and thus the need for high spatial resolution X-ray diffraction. With the recent availability of bright third generation synchrotron sources and progress in X-ray focusing optics, it is now practical to develop an X-ray microdiffraction technique and apply it to characterize materials at such small scales.

© The Author(s) 2015
A.S. Budiman, *Probing Crystal Plasticity at the Nanoscales*,
SpringerBriefs in Applied Sciences and Technology,
DOI 10.1007/978-981-287-335-4_2

15

2.2 Beamline Components and Layout

The Advanced Light Source (ALS) at the Ernest Orlando Lawrence Berkeley National Laboratory in Berkeley, CA, is a third-generation synchrotron radiation source. It is well known as one of the brightest available sources of extreme ultraviolet and soft X-ray radiation in the world. A wide range of scientific activity, ranging from protein crystallography to semiconductor physics, all the way to pioneering technology development of the Extreme Ultra Violet (EUV) lithography technique critical to the continuing scaling of the microelectronics chips have been supported by ALS. In order to provide a wide range of energy spectra, the ALS uses bend magnet, insertion device, as well as superbend sources. The X-ray micro-diffraction beamline described here is located at bend magnet beamline 12.3.2 of the ALS. The beamline provides an extremely bright X-ray beam with a spectral range of approximately 5–22 keV.

Figure 2.1 shows the schematic layout of the X-ray microdiffraction beamline. The X-ray beam from a bending magnet source (1.9 GeV, 400 mA, 250 μm FWHM × 40 μm FWHM, up to 3 × 0.2 mrad divergence in horizontal and vertical respectively) is 1:1 refocused at the entrance of the hutch by a 700 mm long platinum coated silicon toroidal mirror operating at a grazing angle of 5.4 mrad [1, 2]. Among few suitable methods for focusing high brightness white-beam X-rays, Kirkpatrick-Baez (KB) mirror pairs has been chosen for producing our X-ray focused beam in the beamline 12.3.2 as it is the only focusing solution to combine both achromaticity and high efficiency. The principle of these KB mirrors has been described in detail elsewhere [3]. This KB mirrors focus the X-ray syn-chrotron beam into a submicron spot size (0.8 × 0.8 μm FWHM). Acting as an adjustable size source for the KB demagnifying optics inside the hutch are the water-cooled tungsten slits at the entrance of the hutch. In this way spot size can be traded for flux. As white-beam has been mostly used in the studies described in the following chapters, the 4-crystal monochromator in Fig. 2.1 was not in use.

As we can see from Fig. 2.1, after the X-ray white-beam is focused by the KB mirrors, there remain two main components: the sample stage, and the large area

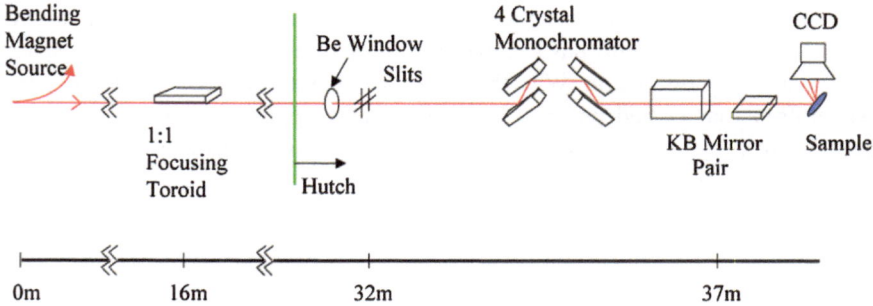

Fig. 2.1 Schematic layout of the beamline 12.3.2 at the advanced light source (ALS), Ernest Orlando Lawrence Berkeley National Laboratory (courtesy of Valek [2])

(a)

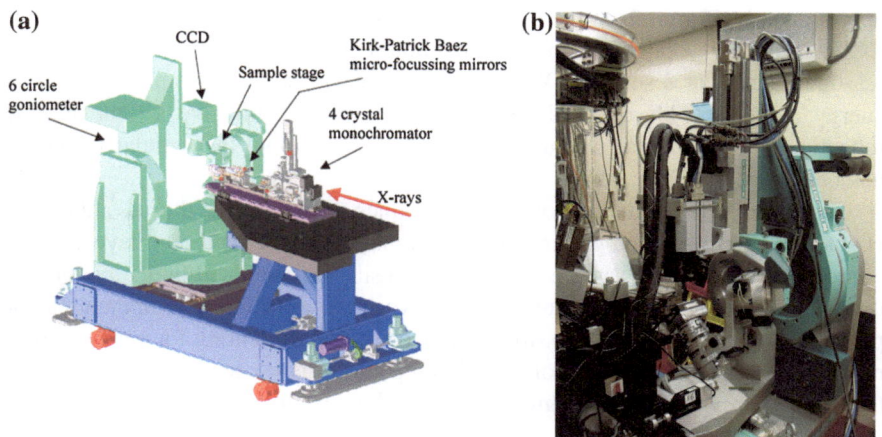

(b)

Fig. 2.2 Experimental endstation for beamline 12.3.2 at the ALS: **a** the schematic of the engineering model, and **b** picture of the actual endstation (courtesy of Tamura et al. [1] and Valek [2])

Fig. 2.3 Side view of our typical experimental setup with the 2-D CCD detector on top and the sample mounted in a 45° reflective geometry; X-ray incoming beam and diffracted beams are shown as white arrows, and the sample movement is precisely controlled by the piezoelectric stage (courtesy of Valek [2])

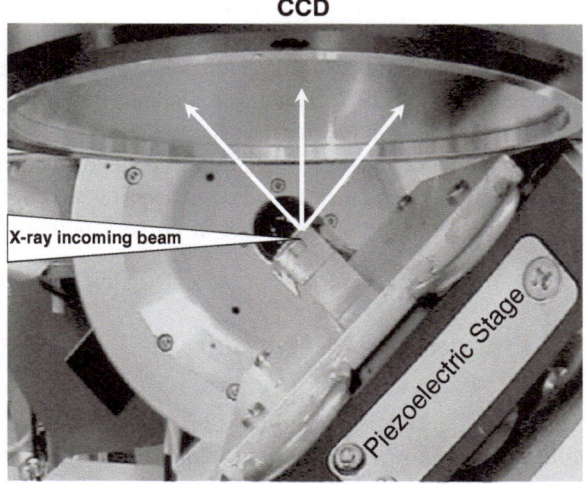

CCD X-ray detector, both of which are mounted on a goniometer as shown in Fig. 2.2. The sample under consideration would sit on a fine XY piezoelectric stage (range of +/−50 μm), which is mounted on a coarse XYZ Huber stage (range of +/−5 mm in XY and +/−10 mm in Z) as shown in Fig. 2.3. The sample can also be mounted on a heating stage for experiments requiring high temperature up to 600 °C.

The diffraction patterns are collected with a MAR133 X-ray CCD (active area of 133 × 133 mm). The sample is usually mounted in a 45° reflective geometry as illustrated in Fig. 2.3, with the CCD detector on a vertical slide at a distance of approximately 50 mm from the sample area illuminated by the beam. This sample-CCD distance is optimized to give maximum total number of reflections at a reasonable angular resolution. When illuminated with a white beam of 5–22 keV energy range, a (111) oriented Al grain, for example, will give a total of ~ 15 reflections, at an angular resolution of 0.01° (details about the angular resolution has been given by MacDowell et al. [3]).

Compared to electron microscopy techniques, X-rays offer the advantages of characterization of buried grains under overlying cap layers and multilayered films without the need of any sample preparation. In situ measurement under a variety of different conditions (in air, liquid, gas and vacuum, at different temperatures and pressures) thus becomes a possibility and opens opportunities for many experimental applications. The lack of sample preparation is especially important since the sample stress state can be greatly affected by any preparation processes.

2.3 Scanning White-Beam X-ray Microdiffraction

White beam Laue diffraction is a standard crystallographic method used to determine crystal orientation without rotation of the sample. However, Laue diffraction is rarely used to measure strain because the precision of most Laue instruments is low compared to modern diffractometers, and because the unit cell volume cannot be determined with a standard Laue measurement. Nevertheless, with suitable instrumentation, such as used in the experimental setting described above, precise determination of crystal orientation and distortional strain is possible. Laue diffraction might even be extended by measuring the energy of one or more reflections to determine the full strain tensor in polycrystalline samples.

2.3.1 Experimental Procedure

The technique at the beamline 12.3.2 that we have used for experimental data collection is known as Scanning white-beam X-ray microdiffraction (μSXRD). As implied by the name, this technique does not use the monochromatic beam capabilities of the beamline and has significant differences with traditional X-ray diffraction techniques. Instead of the angular scans normally associated with X-ray diffraction, the sample is fixed in the μSXRD setup (the crystal orientation and information are captured from its white-beam diffraction image). The "scanning" in the name of the technique refers to the manner that we translate the sample in discrete steps in X and Y axes of the sample stage, within the focal plane of the white radiation microbeam and collect the resulting Laue patterns on a CCD frame. This basic illustration of this experiment is shown in Fig. 2.4.

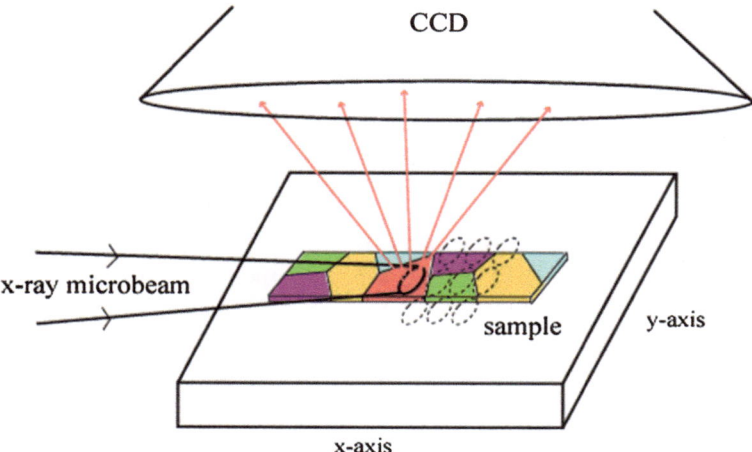

Fig. 2.4 Schematic diagram of a scanning white beam X-ray microdiffraction experiment; Incoming X-ray beam illuminate a volume that may consist of more than one crystal, the diffracted beams from each of the crystal are captured by the CCD detector placed above the sample. Laue patterns are collected from each discrete volume of the sample, followed by the next volume in a discrete-step scanning mode (courtesy of Valek [2])

Any crystal that lies within the illuminated volume will produce a Laue diffraction pattern. A typical white-beam Laue image captured by the CCD X-ray detector, containing of multiple Laue diffraction patterns (from an evidently polycrystalline sample), is given in Fig. 2.5. The Laue patterns can then be analyzed to give information about the complete orientation of individual grains, sample grain structure, deviatoric stress/strain tensors, information on subgrain structures, and crystal deformation/evolution. The methods for analyzing these Laue patterns will be discussed in the coming Sect. 2.3.2.

Prior to that, a rough elemental map of the sample has first to be created using scanning X-ray fluorescence, so that the region of interest can be more precisely pinpointed. For the samples in our experiments, we typically used copper, titanium, platinum and gold fluorescence signals. Subsequently, the area of interest can then be precisely located by using the piezoelectric positioning stage. A typical big-picture elemental map is shown in Fig. 2.6, from which then the exact coordinates to do more precise X-ray microdiffraction scan can be determined, and Laue patterns can thus be collected from each discrete area in the coordinates of interest.

Further references about the experimental approach that we have used could be obtained in a complete manner elsewhere [2].

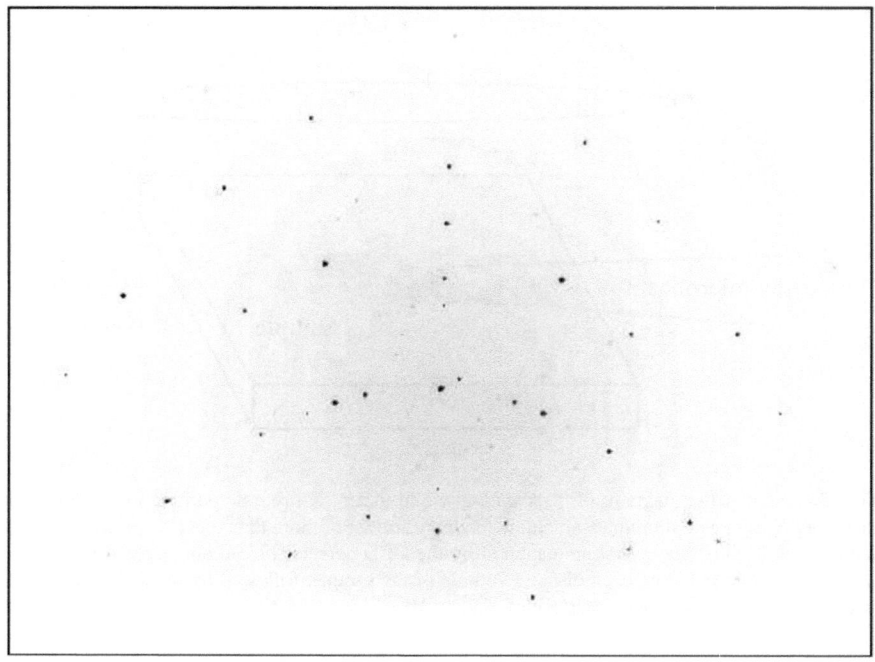

Fig. 2.5 Typical CCD image containing Laue diffraction patterns from copper grains and the underlying silicon substrate; The Si Laue reflections are much brighter than the copper reflections due to silicon substrate's large crystal size relative to copper grains

2.3.2 Data Analysis Using XMAS

A single white-beam Laue diffraction pattern contains a wealth of information about crystallographic orientation, stress/strain and dislocation structure/density in each individual crystal of the typically polycrystalline samples. A typical X-ray micro-diffraction scanning of a sample usually yields hundreds to a few thousands of these Laue diffraction patterns. Without a set of software tools that can rapidly analyze the multiple Laue patterns contained in each of the CCD images produced by our experiments, the abundant raw data could not translate to meaningful information about the materials. A computer automated technique developed for crystallo-graphic indexing, orientation, and strain determinations of grains in thin film samples [4–7] has been used, and a specific custom-made software has been developed in the Beamline 12.3.2 by Dr. Nobumichi Tamura of the Advanced Light Source (ALS), Berkeley Lab, and is called XMAS (X-ray Microdiffraction Analysis Software). It is based on an algorithm first described by Chung and Ice [4] as outlined in Fig. 2.7.

The rest of this Sect. 2.3.2 is organized following through the above flow chart, with the exception of energy measurement for the absolute unit cell parameter calculation, which is outside the scope of this book.

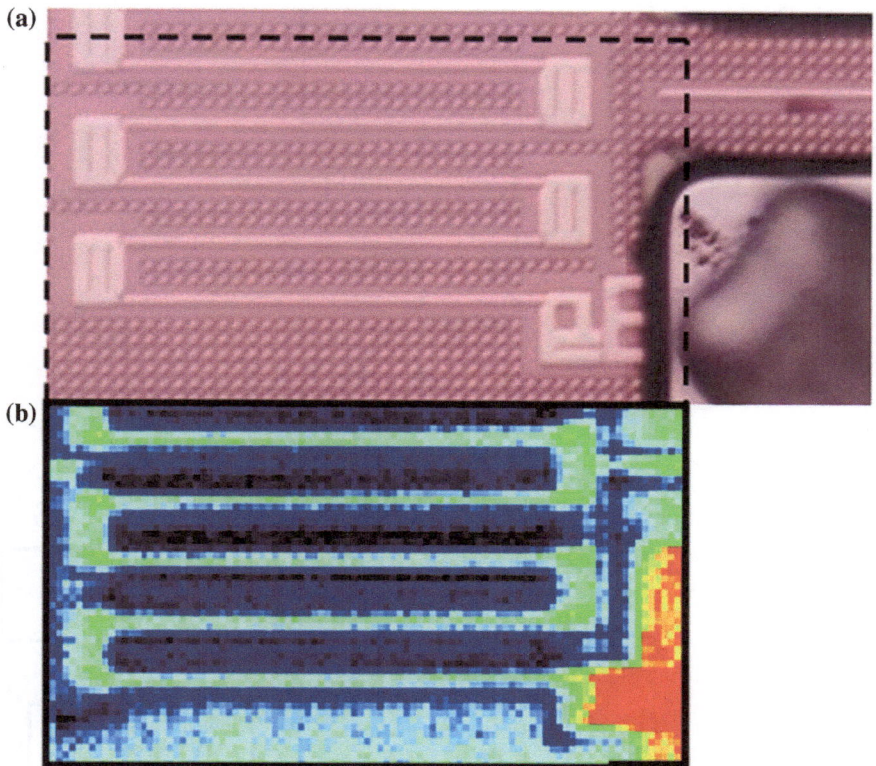

Fig. 2.6 Fluorescence mapping for locating the sample's overall picture; For example **a** shows a series of horizontal copper interconnect lines, and **b** shows the corresponding fluorescence mapping using copper's characteristic radiation wavelength (*red* high intensity of copper fluorescence characteristic radiation, *green* medium, *blue* low). A 0.5 µm step size in *x* and *y* was used

The three key steps for white beam Laue analysis, thus, are:

(a) Laue peak searching and indexation
(b) Crystal unit cell parameter determination
(c) Strain/stress tensor calculation

2.3.2.1 Laue Peak Searching and Indexation

In order to index a Laue pattern, the first step we need to do is to process the image file as captured by the CCD detector. Figure 2.8a shows CCD image collected from a Cu thin film on a single crystal silicon substrate. The XMAS software reads the CCD file and automatically finds peaks using a peak-searching routine. The peaks are classified by their integrated intensity and fit with a 2D Gaussian, Lorentzian, or

Fig. 2.7 Flow chart for Laue reflection data analysis/ processing as followed in the XMAS program used in the Beamline 12.3.2. at the ALS, Berkeley Lab (courtesy of Chung and Ice [4])

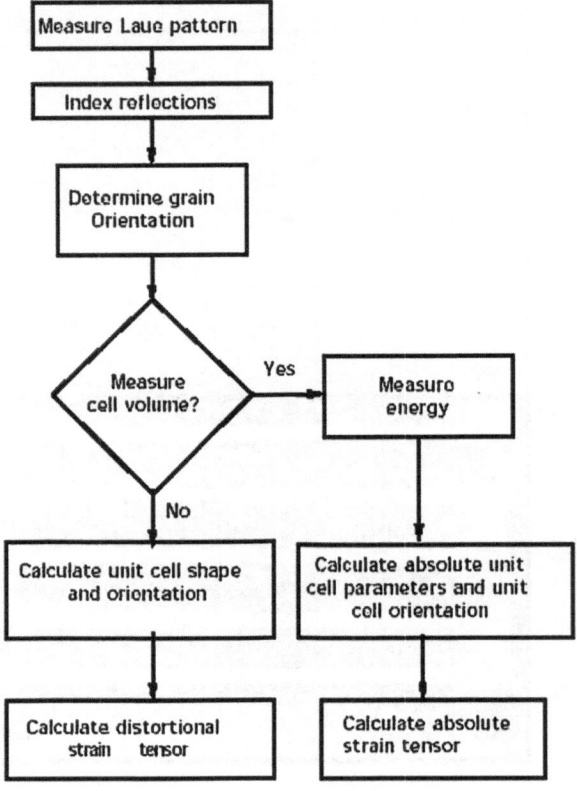

Pearson VII function, which allows the instrument to achieve subpixel resolution on the position of the peaks on the CCD. This is because each peak illuminates several pixels on the CCD due to the divergence of the beam and long tails of its Lorentzian profile. It has indeed been found that the best fit is usually given by a 2D Lorentzian function. The accuracy of this peak position fitting is one of the key limiting factors for strain accuracy measurement of this technique, thus the importance of having highly, precisely calibrated geometrical setup of the instrumentation.

Once the peak positions are established, we can index the Laue pattern with Miller indices. An indexed Cu Laue pattern is shown in Fig. 2.8b. Using the instrument geometry calibration procedure described below, we can calculate the scattering vector attached to each Laue reflection found during the peak fitting routine. The vector k_{in} describes the direction of the incident beam, while a vector k_{out} describes the outgoing diffracted beam directions. If we assume elastic scattering ($|k_{in}| = |k_{out}|$), then q_{exp} describes the directions of the experimental scattering vectors:

$$q_{exp} = k_{out} - k_{in}. \tag{2.1}$$

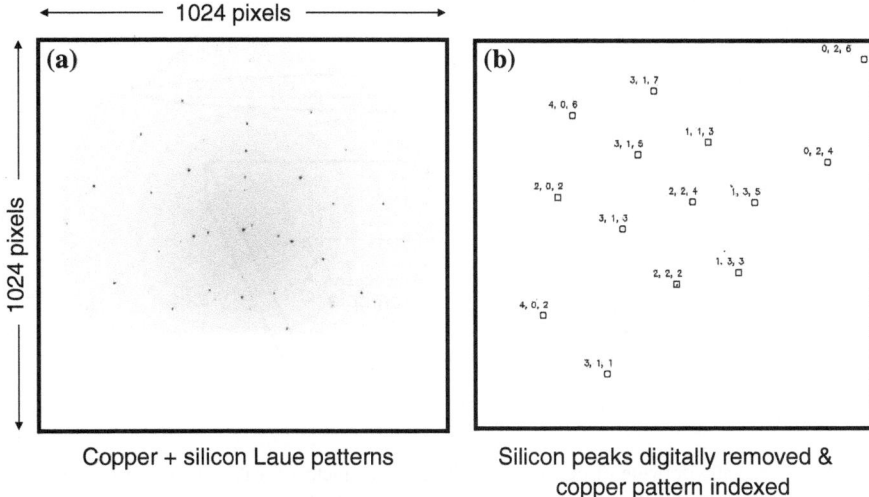

Fig. 2.8 Laue diffraction patterns are indexed by comparing the angles between triplets of experimental diffraction vectors with a list of theoretical angles for the crystal structure; **a** a CCD image consisting of bright silicon peaks as well as copper peaks, **b** indexed copper peaks belonging to a single grain of Cu (there may be other grain/crystal) after the bright silicon peaks are digitally removed

To index the pattern, the code uses a subset N of the brightest Laue reflections. The angles between these q_{exp} vectors for this set of N reflections are tabulated in an 'experimental' list of (N(N − 1)/2) items. This experimental list is then compared to a reference list of angles between the theoretical scattering vectors that have been computed for the crystal structure of interest. The number of theoretical scattering vectors is limited by the energy bandpass of the polychromatic beam. To index the peaks, the code looks for angular matches between triplets of reflections within an adjustable angular tolerance. For each triplet and their corresponding (hkl) indices, the code calculates a complete list of reflections that should be visible on the CCD based on the considered energy bandpass, CCD dimensions, and geometry. The best match is the triplet that matches the largest number of experimental reflections. To limit the number of possible matches, many considerations such as structure factor and crystal symmetry are used.

In the indexing algorithm, the value of N is important. For instance, for a Laue pattern taken on a single-crystal region, a small value of N is used because the most intense reflections will all originate from the same grain. However in the case of a pattern taken in a polycrystalline region, this N value should be increased such that the algorithm can index overlapping Laue patterns. This ability to index multiple Laue patterns in a single CCD frame is key in analyzing microdiffraction data with incident X-ray spot size on the order of the grain size. More than ten overlapping grains in a single CCD frame can be differentiated and indexed with this algorithm.

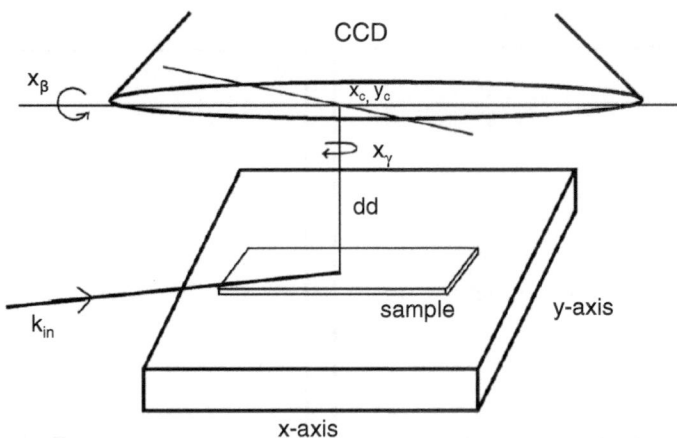

Fig. 2.9 Schematic drawing of the spatial calibration parameters for the CCD detector relative to the sample. These parameters include the sample to detector distance (*dd*), the pitch (x_β) and the yaw (x_γ) of the CCD, and the pixel coordinates of the CCD center channel (x_c, y_c). The vector $\mathbf{k}_{in}$ represents the direction of the incident X-ray beam (courtesy of Valek [2])

Once we know the positions of the reflections on the CCD, we can calculate their Bragg angle, θ, if the exact geometry of the instrumentation systems is defined with high precision (i.e. if the exact position of the CCD with respect to the incoming beam and the point of "impact" on the sample are known precisely). The number of the independent geometrical parameters is five: the X and Y "center channel" coordinates of the CCD, the distance between the CCD center channel and "point of impact" of the incident X-ray beam on the sample, and the two angles describing the tilt orientation of the detector relative to the incident beam (the pitch, x_β, and the yaw, x_γ). The center channel is defined as the X and Y pixel coordinates (x_c and y_c) of the position on the CCD at which the distance between the CCD and point of impact of the incident beam on the sample is minimized. The parameters are shown schematically in Fig. 2.9.

There are two methods can be used to refine the geometrical parameters. The first, triangulation, requires several images are collected at different CCD distances from the sample. Ray-tracing the reflective beams back to the origin is then used to define the geometrical parameters. Moving the CCD, however, introduces a sixth parameter that must be used to describe the angular deviation of the CCD translation stage with respect to the incident beam, which adds additional undesirable complications and errors. The second technique collects a Laue pattern from a calibration sample and uses a non-linear least square refinement to define the necessary geometrical parameters. The ideal calibration sample should be an unstrained large single crystal with no defects (short extinction length) and have a lattice parameter that produces a large enough number of reflections for the energy bandpass of the incident beam. For instance, in the typical thin film materials studies, the reflections from the relatively unstrained silicon wafer substrate can be used as calibration. We can then refine these geometrical parameters by minimizing the function

$$\alpha_0 = \frac{\sum_i w_i (\alpha_i^{th} - \alpha_i^{exp})^2}{\sum_i w_i}, \tag{2.2}$$

where α_i^{th} and α_i^{exp} are the differences in angle between pairs of the theoretical and experimental scattering vectors, respectively, and w_i is a weighting factor. This minimization of course presupposes that the reference Laue pattern is indexed properly and sums over all pairs of reflections visible on the CCD.

2.3.2.2   Crystal Unit Cell Parameter Determination

In the case of strained crystal where unit-cell parameters are unknown, it is still possible to determine the orientation and distortion strain of the unit cell if four independent broad-bandpass reflections are observed. Here we seek to find a matrix U that simultaneously satisfies four observed normal directions. Matrix U is a 3×3 matrix that transforms experimental scattering vectors in the reciprocal space (q_{hkl}) into a real space laboratory coordinate system (q_{xyz}).

$$q_{xyz} = U q_{hkl}. \tag{2.3}$$

Hence, matrix U defines a crystal system, and its 9 unknowns are the parameters of the crystal unit cell (3 crystal rotations, 6 unit cell parameters). They are basically the unit cell parameters $a, b, c, \alpha, \beta, \gamma$ such as shown in Fig. 2.10, and the 3 crystal rotations.

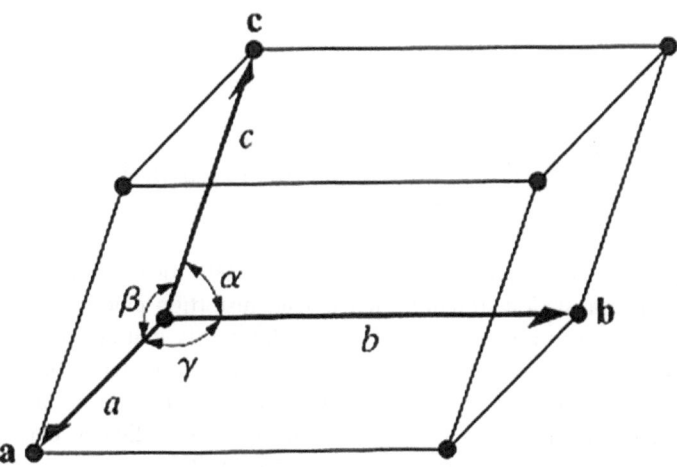

Fig. 2.10 The basic parameters of a unit cell: $a, b, c, \alpha, \beta, \gamma$

With four observed normal directions of scattering vectors, each normal has two direction angles, so altogether there are a total of eight independent equations to solve for the 9 unknowns. However, since the dilatation term cannot be determined without an energy measurement, we restrict ourselves to the determination of the orientation and distortional strain. If dilatation is ignored, there are, therefore, a total of 8 unknowns about the crystal orientation and the unit cell parameters. The one missing unknown is the absolute value of lattice parameters, and thus only the relative magnitude between a, b, and c can be identified, i.e. b/a and c/a.

We can then calculate the unit cell parameters for a single crystalline grain using the same minimization procedure used to calibrate the geometrical parameters of the instrumentation system. Similarly, the differences in angle between pairs of experimental and theoretical scattering vectors are minimized—in this case, Eq. 2.2 is used to calculate the lattice parameter ratios (b/a, c/a) and lattice angles, rather than the geometrical parameters. The instrumentation geometrical parameters (x_c, y_c, dd, x_β, and x_γ) have been fixed from the earlier calibration step.

2.3.2.3 Strain/Stress Tensor Calculation

Upon finding the relative lattice parameters a, b, and c and the lattice angles α, β, and γ, we can use this information to calculate the strain tensor. Here, we follow the procedure outlined by Chung and Ice [4]. Consider for example a unit cell with lattice parameters a_i and α_i, and a Cartesian coordinate system u_i attached to the crystal, such that a_i and u_i are coincident, a_2 is in the u_1u_2 plane and u_3 is perpendicular to the u_1u_2 plane, as modeled in Fig. 2.11.

We specify a position in the crystal using either a vector in the crystal coordinates, v_a, or a vector in the Cartesian coordinates, v_u. To transform from unit cell coordinates to Cartesian coordinates, we use the equation

$$v_u = A v_a, \tag{2.4}$$

where

$$A = \begin{pmatrix} a_1 & a_2 \cos \alpha_3 & a_3 \cos \alpha_2 \\ 0 & a_2 \sin \alpha_3 & -a_3 \sin \alpha_2 \cos \beta_1 \\ 0 & 0 & 1/b_3 \end{pmatrix}. \tag{2.5}$$

Here the b_i and β_i are the reciprocal lattice parameters for the unit cell.

Rigid body translation, rotation, and crystal distortion can transform a vector position v attached to a crystal. Let A_{meas} be the matrix that converts a measured vector v into the measured crystal Cartesian coordinates, calculated from the refined lattice parameters found using Eq. 2.5. Let A_o be the matrix for an unstrained unit cell that converts v into the measured Cartesian crystal coordinates, calculated from the unstrained crystal lattice parameters and angles. Within this definition, there is a transformation matrix T that maps unstrained vectors to strained vectors:

Fig. 2.11 A cartesian coordinate system (u_i) attached to a real space unit cell (a_i, a_i). Note that we define $u_1 = a_1/|a_1|$; $u_3 = a_1 \times a_2/|a_1 \times a_2|$; and $u_2 = a_3 \times a_1 \times |a_3 \times a_1|$ (based on Chung and Ice [4], figure courtesy of Valek [2])

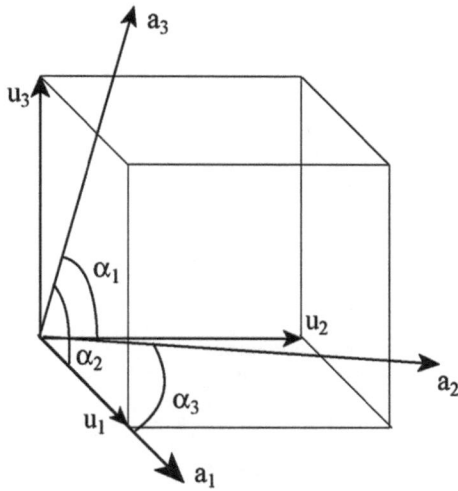

$$A_{meas} = TA_o. \tag{2.6}$$

T can contain both distortional components and rigid body rotation terms. Rigid body translation has been eliminated by defining the crystal and Cartesian coordinate systems to have a common origin. By using the definition of the strain tensor, ε_{ij}:

$$\varepsilon'_{ij} = (T_{ij} + T_{ji})/2 - I_{ij} \tag{2.7}$$

where I is the identity matrix, we can thus eliminate the antisymmetric rotational component.

Finally, the deviatoric stress tensor can be derived from the deviatoric strain tensor by applying the anisotropic stiffness coefficients C_{ijkl} for the material:

$$\sigma'_{ij} = C_{ijkl}\varepsilon'_{kl} \tag{2.8}$$

We have described the data analysis process behind the XMAS program which will be used heavily in the following chapters. Overlapping Laue patterns from multiple grains can be indexed by an iterative process which compares possible indices for several reflections with the observed angles between the reflections. Once the reflections are indexed, the unit cell parameters and crystal orientations can be determined. From the deviation between the observed Laue pattern and that of the unstrained crystal, deviatoric strain tensor, and thus deviatoric stress tensor can be calculated. These methods allow for rapid data analysis which makes mechanical probing of polycrystalline materials practical.

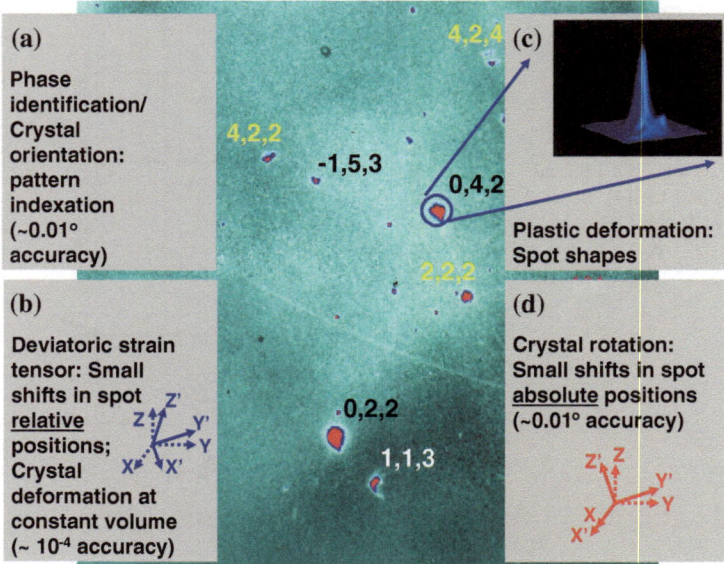

Fig. 2.12 A single white-beam CCD image consisting of multiple sets of Laue diffraction peaks (three colors: *black*, *yellow* and *white*) from a Cu polycrystalline sample; The determination of **a** and **b** have been described in Sect. 2.3, while the determination of **c** and **d** will be discussed in this section

2.4 Local Plasticity Probing Using Whitebeam µXRD

The determination of the crystallographic orientation and stress/strain (deviatoric) in each of the individual crystal has been discussed (Sect. 2.3.2). It is basically done using the position and change in the relative position of the Laue diffraction peaks on the CCD image. Having done extensive experiments using the technique [8–18], we now realize another source of wealth of information that the technique gives is the shape of the Laue diffraction peak itself, as well as the change in the absolute position of the peak. Figure 2.12 shows features in Laue diffraction patterns and their associated structural and mechanical information that they contain about the material/sample.

First, the shape of the Laue peak itself provides information about plastic deformation in the crystal, especially the one involving the Geometrically Necessary Dislocations (GNDs), and then secondly, the change in the absolute position of the Laue peak gives us the rotational deformation of the crystal body. This Section will describe and discuss about the data analysis procedure for the determination of plastic and rotational deformation in crystals that will be used again and again throughout the chapters in this book.

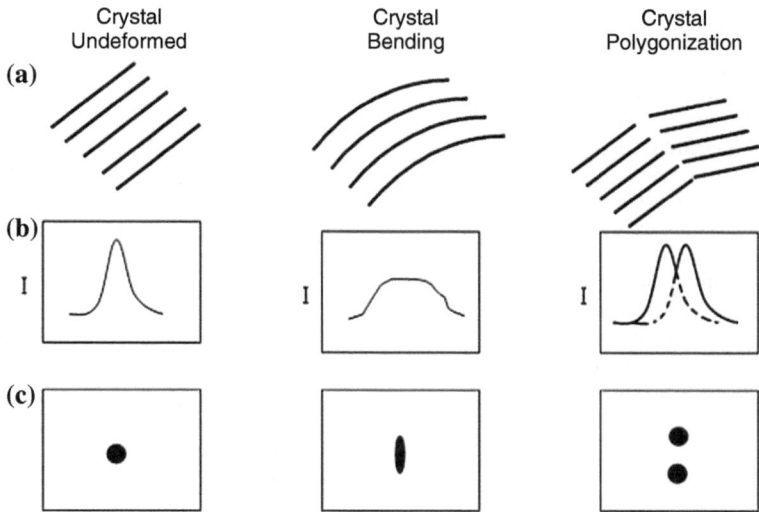

Fig. 2.13 Schematic diagrams comparing a set of crystal planes in their undeformed, bent/curved and polygonized states (**a**), while their expected Laue diffraction peaks corresponding to each of the crystal states are shown in (**b**), in intensity scanning over angle consisting of the Bragg angle/s, and (**c**), in CCD detector space (courtesy of Spolenak et al. [11])

2.4.1 Crystal Bending, Polygonization and Rotation

In traditional X-ray Diffraction (XRD) experiments, a peak at a certain θ angle means a particular (hkl) plane is detected, i.e. the particular θ angle, d_{hkl} interplanar distance and $\lambda_{Cu,k\alpha}$ wavelength (for instance) conspire such as to satisfy Bragg's Law:

$$n\lambda_{Cu,k\alpha} = 2d_{hkl}\sin\theta. \tag{2.9}$$

Using this methodology, only elastic deformation of crystal can be determined, where strain can simply be computed from the difference of interplanar distances between atomic planes, before and after the deformation. In contrast, plastic deformation may involve curved and polygonized crystal planes, such as shown in Fig. 2.13. In such configuration, a particular crystal plane (with a fixed d_{hkl} interplanar distance) may have a range of values of θ angle. This lattice curvature information will be lost in a traditional XRD (using a single wavelength and no tilt). However, using a white-beam X-ray, which also has a range of wavelengths, Bragg's law can still be satisfied even though the θ angle varies. This means that with white-beam, a presence of lattice curvature can still be detected and even measured.

In the case of a curved crystal plane (crystal bending), a broadened Laue diffraction peak in a certain direction would be observed (Fig. 2.13) instead of a single, sharp, rounded peak typical of an undeformed crystal plane. The Laue peak

Fig. 2.14 Schematic diagrams comparing two bodies of crystal in their undeformed/unrotated and rotated states (**a**), while their expected Laue diffraction peaks corresponding to each states are shown in (**b**), in intensity scanning over angle consisting of the Bragg angle (there is a shift in the absolute angular position of the peak), and (**c**), in CCD detector space (again, a shift in the position of the peak on the CCD detector space)

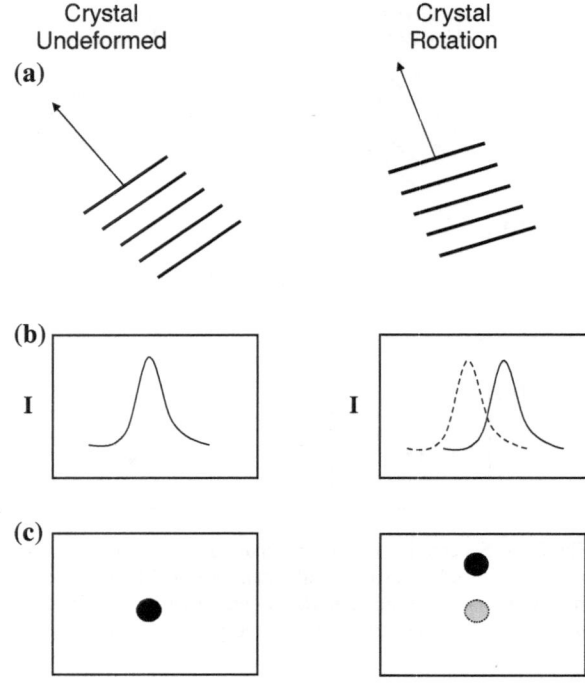

broadening (also called "peak streaking") is continuous representing the continuum of θ values involved in the curved crystal. Crystal polygonization, in contrast, involves a multiple, discreet peaks coming from slightly misoriented sub-crystal structures (such as shown, again, in Fig. 2.13). Such polygonization often occurs following bending of crystal where more and more dislocations are piling up, and thus become unstable with respect to glide and eventually climb to form a low-angle boundary [19–22].

This is the unique capability of our white-beam X-ray diffraction technique. Using this white-beam methodology, we can start to directly and quantitatively study the plastic deformation in crystals. This technique is especially sensitive towards the crystal bending or polygonization in their simplest geometries, where a net density of parallel same-sign Geometrically Necessary Dislocations (GND's) is formed.

Furthermore, in traditional XRD, when a body of crystal rotates, the θ angle changes and without rotating the sample, Bragg's law (Eq. 2.9) can no longer be satisfied. In contrast, using a white-beam X-ray, another wavelength would simply be available to satisfy diffraction condition, and thus would give a constructive interference in the diffracted intensity (a Laue diffraction peak) instead on a slightly different position in the CCD camera, such as shown in Fig. 2.14. In other words, a crystal rotation would appear as a shift in the absolute positions of each of the Laue peaks belonging to the same crystal (it is to be noted that a change in the *relative* position of the Laue peaks—relative to other peaks belonging to the same crystal—in contrast, does not mean crystal rotation, but instead, shear deformation or in other words, deviatoric strains, as has been discussed in the Sect. 2.3).

2.4.2 Quantitative Peak Study

Our technique of White-beam X-ray Microdiffraction can detect and measure the curvature of the lattice, or in the case of polygonized subgrain structure, the low-angle subgrain boundary. From how much the crystal under consideration curved, or bent, or polygonized, we can obtain quantitative data related to the plastic deformation the crystal must have undergone. The width of a Laue diffraction peak contains information on the dislocation density within a grain and the splitting of Laue diffraction peak into multiple individual spots can be used to measure the formation of low angle grain boundaries.

The X-ray Microdiffraction Analysis Software (XMAS) program allows us to select an individual Laue diffraction peak from a CCD image and study it quantitatively. The peak can be plotted in 2-D or 3-D using the detector coordinates. More meaningfully, as far as the crystal is concerned, is the ability to convert the peak into either theta-chi (θ-χ) space or reciprocal space. In θ-χ space, the Bragg angle, θ, can be calculated using the equation:

$$\theta = \frac{\cos^{-1}[k_{in} \cdot k_{out}]}{2}, \tag{2.10}$$

where k_{in} and k_{out} re the incident beam and diffracted beam. The χ angle is defined as the angle of the diffracted beam within the plane perpendicular to the incident beam. The χ angle can be found by projecting the diffracted beam into the XZ_{lab} plane, and then finding its angle relative to the Z_{lab} axis. The distortion of the θ-χ space on the CCD frame typical for our experimental geometry is shown in Fig. 2.15. The XMAS program allows conversion to remove the distortion caused by using a flat CCD to detect the Laue patterns in three dimensional space.

Once the peaks are in θ-χ coordinates, we have the real rotation of the crystal planes associated with the plastic deformation. This angular change in the geometry of our crystals can thus be further converted into the change in the geometrically necessary dislocation structures (plasticity) that the crystal must have undergone during the deformation. For illustration purposes, we use our real experimental results (Fig. 2.16), in the form of Laue peak intensity contour on a θ-χ space, its intensity trace along a certain χ angle, and the Full Width Half Maximum (FWHM) measurements from each of the peak, to give examples of the methodology to determine the extent of plastic deformation from the peak study.

Figure 2.16 shows typical Laue diffraction peaks (in θ-χ coordinates) coming from an undeformed crystal, a plastically bent/curved crystal and a polygonized crystal into two low-angle subgrain structures. Peak broadening and split occur primarily across the theta direction in this particular experiment. The peak

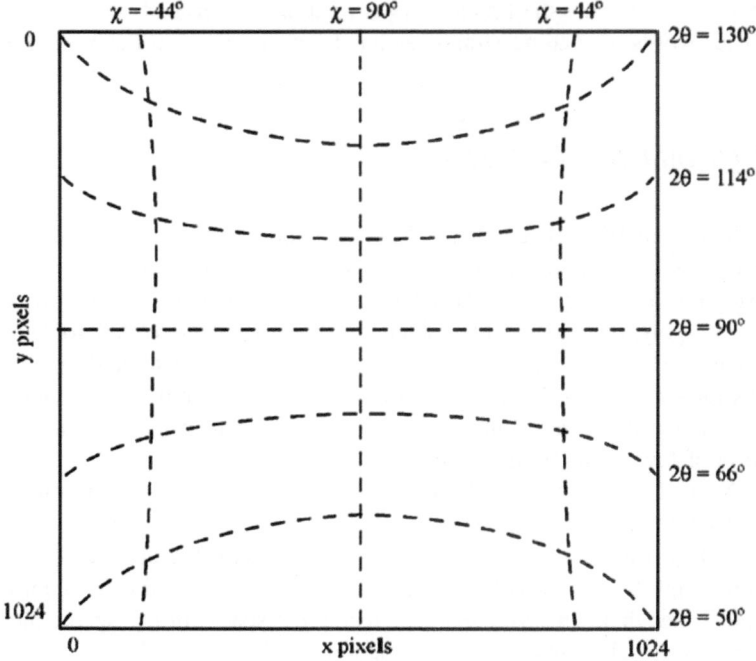

Fig. 2.15 The distortion of 2θ-χ space on the CCD detector typical for our experimental geometry; lines of constant 2θ and constant χ are shown, as well as the orientation of the CCD pixels (courtesy of Valek [2])

broadening can be quantified by defining $\Delta\theta_{FWHM}$ as the difference between the full width at half maximum (FWHM) of the peaks plotted in the θ-χ space along the θ direction at a constant value of χ, chosen to divide the peak in half. These intensity traces are plotted in Fig. 2.16 (in the column next to the Laue peaks) and are subsequently fit using a Lorentzian function as shown in the figure. Clearly from the FWHM measurements, we see peak broadening from (a) to (b) which represents plastic bending of crystal such as illustrated in (d). Similarly, the measurement of the θ angle difference between the two peaks in (c) suggests polygonization of the crystal across the theta direction as illustrated again in (d). Data can also be quantified within the reciprocal space, as an alternative to the θ-χ space.

Knowing the relevant geometrical dimensions of the crystal in the experiment, and the calibrated CCD-detector-to-sample distance, the peak broadening ($\Delta\theta_{FWHM}$) can be converted into the radius of the curvature of the crystal bending, Cahn [23] and Nye [24] further suggested that from the curvature of the crystal bending, we can determine the amount of Geometrically Necessary Dislocations (GND's) that have to be in the crystal to accommodate such a change in the geometrical shape of the crystal:

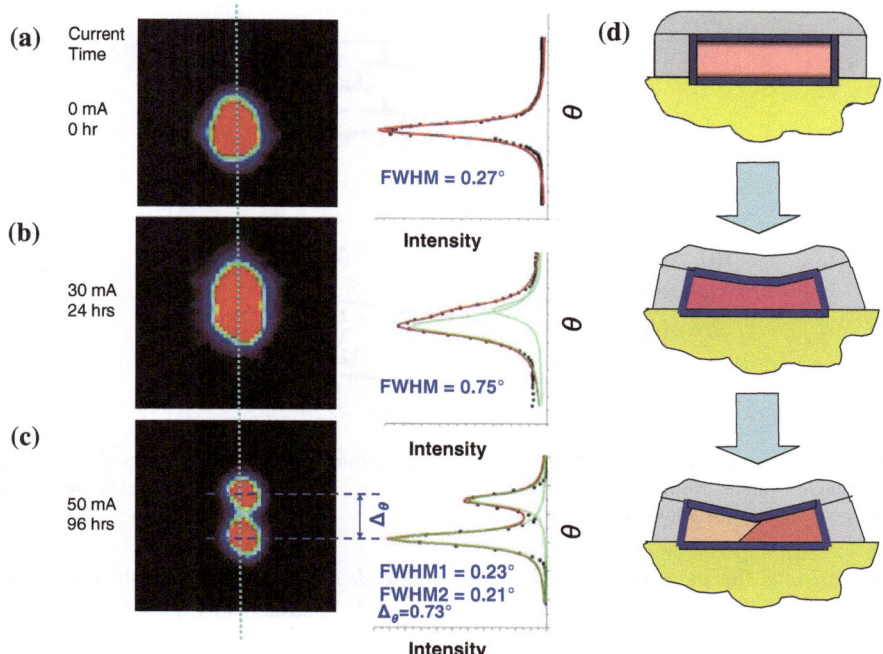

Fig. 2.16 Quantitative measurement of the evolution of a Laue diffraction peak (Peak 002) of a copper grain/crystal in an interconnect line under high current density electrical loading (electromigration) [16], and will be described again in Chap. 3; broadening is shown as the peak evolves from (**a**) to (**b**), while peak splitting, signifying the formation of low-angle grain boundaries, is evident in (**c**); (**d**) is an illustration of the evolution of the grain/crystal that the Laue peak evolution indicates

$$\rho = \frac{1}{Rb}, \tag{2.11}$$

where ρ is the dislocation density of GND's, R is the radius of curvature, and b is the component of the Burgers vector on the neutral axis. This is illustrated in Fig. 2.17a.

The peak split intensity profiles (Fig. 2.16) show clearly resolvable peaks indicating that the initial grain (a) breaks up into crystal segments with their own well-defined orientation (c). From the splitting of the Laue peaks, Δ_θ, the misorientation between adjacent subgrains can be measured. Using the measured misorientation, the spacing between dislocations in the boundary can be derived using the Burgers relation for a small angle grain boundary:

$$\Delta_\theta = \frac{b}{L}, \tag{2.12}$$

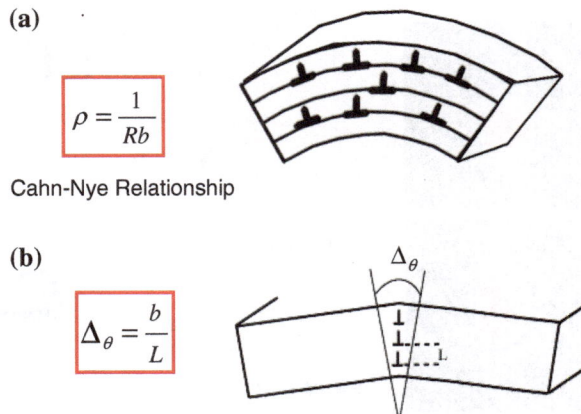

Fig. 2.17 Determinations of densities of GND's from geometrical shape of the crystals; **a** crystal bending/curvature, using Cahn-Nye [23, 24] relationship, and **b** crystal polygonization, using Burgers relation for a low-angle grain boundary

where Δ_θ is the measured misorientation angle, b is the Burgers vector and L the spacing between dislocations in the boundary, as illustrated in Fig. 2.17b.

References

1. Tamura N, MacDowell AA, Spolenak BC et al (2003) Scanning X-ray microdiffraction with submicrometer white beam for strain/stress and orientation mapping in thin films. J Synchrotron Radiat 10:137–143
2. Valek BC (2003) X-ray microdiffraction studies of mechanical behavior and electromigration in thin film structures. Dissertation, Stanford University
3. MacDowell AA, Celestre RS, Tamura N et al (2001) Submicron X-ray diffraction. Nucl Instrum Methods Phys Res 467–468:936–943
4. Chung JS, Ice GE (1999) Automated indexing for texture and strain measurement with broad-bandpass X-ray microbeams. J Appl Phys 86:5249–5255
5. Tamura N, Chung JS, Ice GE et al (1999) Strain and texture in Al-interconnect wires measured by X-ray microbeam diffraction. Mater Res Soc Proc 563:175–180
6. Tamura N, MacDowell AA, Celestre RS et al (2002) High spatial resolution grain orientation and strain mapping in thin films using polychromatic submicron X-ray diffraction. Appl Phys Lett 80:3724–3726
7. Ice GE, Larson BC (2000) 3D X-ray crystal microscope. Adv Eng Mater 2:643–646
8. Valek BC, Bravman JC, Tamura N et al (2002) Electromigration-induced plastic deformation in passivated metal lines. Appl Phys Lett 81:4168–4170
9. Tamura N, Spolenak R, Valek BC, Manceau A, Meier Chang N, Celestre RS, MacDowell AA, Padmore HA, Patel JR (2002) Submicron X-ray diffraction and its applications to problems in materials and environmental science. Rev Sci Instrum 73:1369–1372
10. Valek BC, Tamura N, Spolenak R et al (2003) Early stage of plastic deformation in thin films undergoing electromigration. J Appl Phys 94:3757–3761
11. Spolenak R, Brown WL, Tamura N et al (2003) Local plasticity of Al thin films as revealed by X-ray microdiffraction. Phys Rev Lett 90:096102–096104

12. Budiman AS, Tamura N, Valek BC et al (2004) Materials, technology and reliability for advanced interconnects and low-k dielectrics. Mat Res Soc Proc 812:345–350
13. Wu AT, Tu KN, Lloyd JR et al (2004) Electromigration-induced microstructure evolution in tin studied by synchrotron X-ray microdiffraction. Appl Phys Lett 85:2490–2492
14. Tamura N, Padmore HA, Patel JR (2005) High spatial resolution stress measurements using synchrotron based scanning X-ray microdiffraction with white or monochromatic beam. Mat Sci Eng 399:92–98
15. Wu AT, Tamura N, Lloyd JR et al (2005) Synchrotron X-ray micro-diffraction analysis on microstructure evolution in Sn under electromigration. In: MRS proceedings, vol 863, Cambridge University Press, Cambridge, pp B9–10
16. Budiman AS, Tamura N, Valek BC et al (2006) Crystal plasticity in Cu damascene interconnect lines undergoing electromigration as revealed by synchrotron X-ray microdiffraction. Appl Phys Lett 88:233515
17. Budiman AS, Tamura N, Valek BC et al (2006) Electromigration-induced plastic deformation in Cu damascene interconnect lines as revealed by synchrotron X-ray microdiffraction. In: Materials, technology and reliability of low-k dielectrics and copper interconnects proceedings of symposium F held at the 2006 MRS spring meeting, vol 914, pp 295–304
18. Budiman AS, Hau-Riege CS, Besser PR et al (2007) Plasticity-amplified diffusivity: dislocation cores as fast diffusion paths in Cu interconnects. In: Proceedings of 45th annual IEEE international reliability physics symposium, pp 122–127
19. Cahn RW (1949) Recrystallization of single crystals after plastic bending. J Inst Metals 76:121
20. Gilman JJ (1955) Structure and polygonization of bent zinc monocrystals. Acta Metall 3:277–288
21. WR Hibbard, CG Dunn (1956) Creep and recovery. American Society for Metals, Metals Park, Ohio, p 52
22. Patel JR (1958) Arrangements of dislocations in plastically bent silicon crystals. J Appl Phys 29:170–176
23. Cahn RW (1949) Recrystallization of single crystals after plastic bending. J Inst Met 76:121–141
24. Nye JF (1953) Some geometrical relations in dislocated crystals. Acta Metall 1:153–162

Chapter 3
Electromigration-Induced Plasticity in Cu Interconnects: The Length Scale Dependence

Abstract The early stage of electromigration in Cu interconnects, before the visible structural damage is discussed in this chapter. It is shown, through the μSXRD technique that plastic deformation occurs on the early stage of electromigration. The behavior of the deformation depends on the interconnect width (the crystal grains are mainly rotating in the narrow lines, while crystal bending is observed in the wider lines). Moreover, in the case of wide interconnects, the direction of bending axis is very close to the direction of electrical current.

Keywords Electromigration induced plasticity · Cu interconnects · Size effect · Grain evolution · μXRD · Linewidth effect · Texture

3.1 Introduction

The scanning X-ray submicron diffraction (μ-SXRD) technique using focused synchrotron radiation white beam developed in the Beamline 12.3.2 at the ALS Berkeley Lab has been used to study the microstructural evolution at the granular level of Cu polycrystalline lines during electromigration (EM). An unexpected mode of plastic deformation was observed in damascene Cu interconnect test structures during an in situ electromigration experiment and before the onset of visible microstructural damage (void, hillock formation). We show here, using this synchrotron technique, that the extent of this electromigration-induced plasticity is dependent on the line width. In wide lines, plastic deformation manifests itself as grain bending and the formation of subgrain structures, while only grain rotation is observed in the narrower lines. This EM-induced plasticity tends to occur in large grains spanning across the width of the Cu lines in the form of grain bending and polygonization, whereas smaller grains tend to just rotate.

Furthermore, we observe that the bending axis of this plastic deformation coincides with one of the $\langle 112 \rangle$ line directions of the known slip systems for FCC crystal, and that it is always very close (within a few degrees) to the direction of the

© The Author(s) 2015

A.S. Budiman, *Probing Crystal Plasticity at the Nanoscales*,
SpringerBriefs in Applied Sciences and Technology,
DOI 10.1007/978-981-287-335-4_3

electron flow in the lines. This finding suggests a correlation of the proximity of a $\langle 112 \rangle$ line direction to the direction of electron flow with the occurrence of plastic behavior. This deformation geometry leads us to conclude that dislocations introduced by plastic flow lie predominantly in the direction of electron flow and may provide additional easy paths for the transport of point defects. Since these findings occur long before any observable voids or hillocks are formed, they may have direct bearing on the final failure stages of electromigration.

3.2 Background

There is a great deal of interest in the mechanical behavior of materials especially in today's nano and microscale devices, built near the scale of their microstructural inhomogeneities. Major reliability and device concerns with such nanoscale Cu lines include electromigration [1], interface instability due to thermal stress associated with Joule heating [2], and increased resistivity due to interface and grain boundary scattering. [3] Crystal plasticity and how it progresses during device operation might play an important role. For instance, texture has been known to impact the electromigration performance of the interconnect line. [4] Understanding the exact mechanism and evolution of the plastic deformation in a line with a certain texture therefore might lead us to a texture of higher resistance to EM.

In a related study [5], a very early stage of plastic deformation and microstructural evolution during an EM test was detected in Al(Cu) interconnect lines shown in Fig. 3.1 (courtesy of Valek et al. [5]), long before any macroscopic damage became visible, by using a synchrotron technique involving white-beam X-ray microdiffraction [6]. In addition it was observed that during in situ EM a gradient of plastic deformation evolves along the line, which results in bending and in polygonization of the largest grains between the cathode and the anode end. Smaller grains do not readily deform but do rotate as EM proceeds. Plastic deformation is initiated at the cathode end and gradually progresses toward the anode end while EM is occurring until a steady state is reached.

In the present study, we investigate the evolution of plasticity in Cu interconnect lines during similar electromigration experiments. It is intended to study the generality of the observation of the electromigration-induced plastic deformation to another metal system under similar loading situation. Copper is a reasonable candidate for this study. Furthermore, with the ever decreasing size of interconnect widths to the nanometer range, copper continues to face yet newer/higher level of challenges, and thus fundamental understanding of degradation mechanisms during electromigration in Cu interconnects remains an important topic. In particular, we report here the length scale dependences of the extent of the electromigration-induced plastic deformation. We also correlate the deformation geometry to the occurrence of plastic behavior.

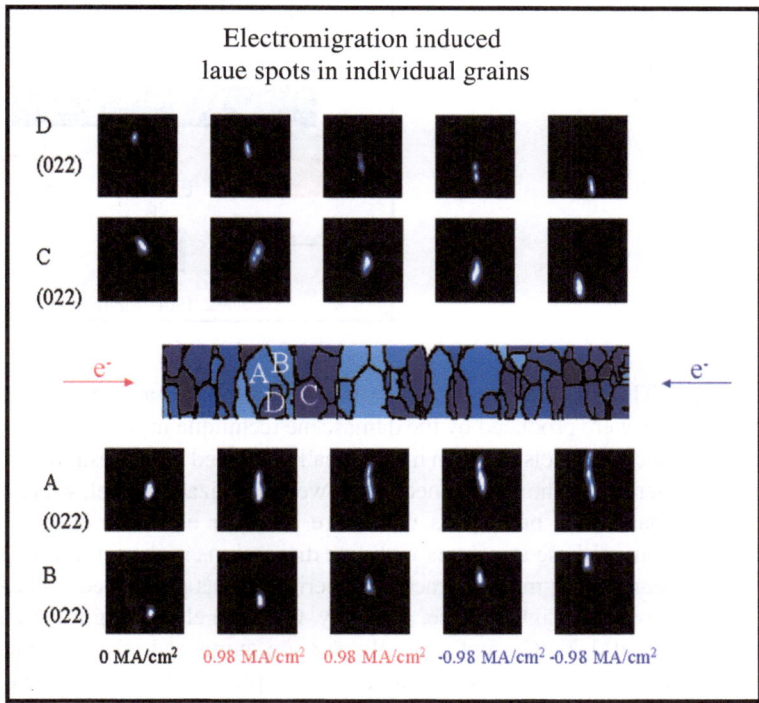

Fig. 3.1 The first observation of the electromigration-induced plastic deformation in Al(Cu) interconnects; [5] the evolution of Laue diffraction spots from grains *A*, *B*, *C* and *D* (locations in the line are as shown in the grain mapping) from initial state to after some electromigration with reversed current directions (courtesy of Valek et al. [5])

The synchrotron technique of scanning white beam X-ray microdiffraction has been described in a complete manner in the previous chapter, and in the literature [6]. Other in situ microstructure characterization studies have given valuable insights on the degradation mechanism of electromigration in Cu lines [7–9]. To complement these studies, the high brilliance synchrotron radiation here allows in situ studies of crystal lattice rotation and its evolution during electromigration. This is an important piece of information that contributes to the fundamental understanding of EM degradation mechanisms.

3.3 Experimental

The interconnect test structure investigated in the present study is an electroplated Cu damascene line manufactured by Intel Corporation. The test line has dimensions of 70 μm in length and approximately 1 μm in thickness, with two different widths of 1.6 μm and 0.6 μm. The lines are embedded in a fluorinated SiO_2/SiOF interlayer

Fig. 3.2 The schematic
diagram of the Cu
electomigration test structures
manufactured by Intel
Corporation

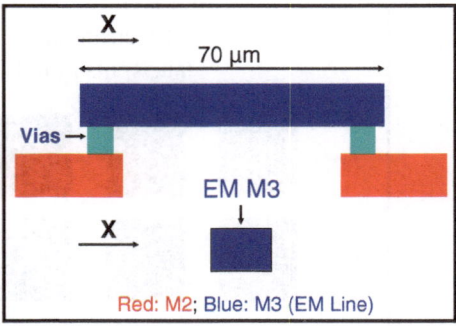

dielectric film. The structure is then passivated with nitride and polyimide. The interconnect lines were produced by the damascene technique in which the copper is plated out into open channels and then mechanically polished to the desired thickness. Both vias at either end of the line connect to a lower metallization level, which in turn connects to unpassivated bond pads which are used for electrical connection. A schematic diagram of these structures with line dimensions is shown in Fig. 3.2.

The white beam X-ray microdiffraction experiment was performed on beamline 12.3.2. at the Advanced Light Source, Berkeley, CA. The electromigration test was conducted first at 300 °C. Current and voltage were monitored at 10 s increments. The sample (width = 1.6 μm) was scanned in 0.5 μm steps, 10 steps across the width of the line and 160 steps along the length of the line, for a total of 1,600 CCD frames collected. A complete set of CCD frames takes about 6–7 h to collect. The exposure time was 4 s plus about 10 s of electronic readout time for each frame. In this manner the Laue pattern and information regarding plastic deformation for each grain in the sample was collected for each time step during the experiment. The current was ramped up to 50 mA ($j = 3.1$ MA/cm^2) over the course of 96 h, then set at that value for the rest of the test.

The second group of tests was conducted at a higher temperature, 360 °C, for reasons that will be discussed later in the chapter. The narrow sample (0.6 μm) was scanned in the same manner as above, except that the current ramp up was up to 20 mA ($j = 3.3$ MA/cm^2) over the course of 96 h, then set at that value for the rest of the test.

3.4 Results and Discussion

3.4.1 Microstructure of the Cu Interconnect Lines

We first describe the microstructure of the wide (1.6 μm) damascene Cu test structures. The grain out-of-plane and in-plane orientations in these lines as determined by white beam X-ray microdiffraction through the passivation layer are shown in Fig. 3.3a and b, respectively.

Fig. 3.3 Grain orientation mapping of the wide (1.6 μm) passivated Cu lines using synchrotron-based X-ray microdiffraction with focused beam (FWHM ∼0.8 μm) **a** crystal ⟨111⟩ out-of-plane orientation; **b** crystal ⟨100⟩ in-plane orientation

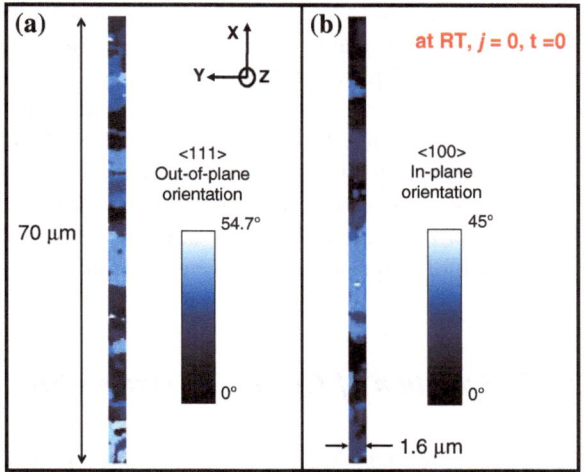

Figure 3.3a shows all grains along the interconnect line with the ⟨111⟩ direction of the individual grains varying from 0° (normal to sample surface) to 54.7° away from the normal of the sample surface. In other words, the darker the blue color of the grains, the closer those grains to having ⟨111⟩ out-of-plane orientation, and the lighter the blue color, the further away from having ⟨111⟩ out-of-plane orientation the grains are. It is evident from Fig. 3.3a that Cu grains in this damascene interconnect test structures were found in a wide range of out-of-plane orientations. This is consistent with what has been observed by some researchers in industry-relevant Cu interconnect test structures. [10–12] Only a few of the grains (black-colored) are actually ⟨111⟩ grains, some others are quite close (dark blue) to being ⟨111⟩ grains, yet many others are far off (the light blue to white-colored), even approaching ⟨100⟩-out-of-plane-oriented grains (54.7° off).

Most of the grains are large grains spanning across the width of the line. It is reasonable to suspect that the Cu grains here extend through the whole thickness of the line, making it a bamboo structure. This is true along the line, however, in areas close to the via regions, grains are smaller, and more likely to be three-dimensional in structure (not bamboo). This trend has again been reported in the literature, [9–13] where the wider Cu damascene lines have been shown to be closer to the behavior of Cu blanket films (bamboo structure), whereas the narrower lines exhibit behaviors toward three-dimensional polycrystalline structure.

Similar observations were found with the in-plane orientation. Figure 3.3b shows all grains along the interconnect line with the ⟨100⟩ direction of the individual grains, projected to the sample surface, varying from 0° (exactly lining up with the positive x direction) to 45° away from x-axis ⟨100⟩. It may appear that the in-plane orientations of the Cu grains in this line are not too widely-ranged. However, we understand that as it is projected ⟨100⟩ directions on the sample surface, the same color in this map does not necessarily mean a single crystal/grain, but in fact, contrasting it to the out-of-plane orientation mapping (Fig. 3.3a), it is evident that the same color in this map can actually be a few grains. Thus, we suspect that the Cu grains in this wide line, are as varied in in-plane orientations, as they are in out-of-plane orientations.

Exact grain orientation mapping of the narrow line (0.6 μm) of these Cu damascene test structures, however, proves to be difficult and rather unreliable. The X-ray spot size (0.8 × 0.8 μm) currently used in ALS beamline 12.3.2 was relatively large for the dimensions of the narrow line (0.6 μm linewidth). That makes diffraction spot indexation often very difficult and thus mapping of grain orientations and other further quantitative analyses unreliable.

3.4.2 Evolution of Cu Grains During Electromigration

Electromigration tests were conducted in situ on the damascene Cu test structures. Figure 3.4 shows the evolution of the Laue diffraction spots for several grains in the wide (1.6 μm) Cu line during the in situ EM experiment. If we examine the individual diffraction spots after electromigration in some detail we find that in certain grains the spots broaden not in any random direction but always in the y direction across the line. The Laue diffraction spots coming from undeformed crystals are nominally rounded in shape, such as shown in Fig. 3.4 at the initial stage of the EM test (at $j = 0$, $t = 0$, and room temperature).

The diffraction spots have been converted to q-space (reciprocal space), with the x-axis along the length of the line, the y-axis across the line, and the z-axis normal to its surface. We find that as the EM test progresses in certain grains the spots broaden, and in some grains, they split into two different spots. This broadening and splitting of diffraction spots is observed not in any random direction but always along the y-axis in q-space, which translates physically to the direction across the Cu line.

Broadening of the peak is observed, in this wide test structure sample, in a few grains that tend to be the large ones and the ones spanning across the width of the line. In one particular such grain, Grain 2 (in the middle of the line), we also show (Fig. 3.5a–c) the results of digital intensity traces across the broadening direction of the diffraction spots in the initial, mid and end-state (after the end of the EM test). We have actually seen this Fig. 3.5 from our earlier discussion about the quantitative peak study methodology from Chap. 2 of this book (there, as Fig. 2.16, it was used to help explain our technique/analysis method).

Such broadening and splitting of the diffraction spots were observed in all three different wide test structure samples examined in our experiments. In each of them, a few grains (between 5 and 9) among a total of usually around 100–150 grains, were found with this observed behavior after similar electromigration test time, current and temperature.

The broadening of the diffraction spots represents crystal bending of the Cu grains in the line, whereas the split diffraction spots indicate the formation of low-angle boundary subgrain structures. During this electromigration test, it is thus

Fig. 3.4 Evolution of Laue diffraction spots (in q-space) of three grains (one at the cathode-end, another in the middle, and the last one at the anode-end of the line) during an in situ EM experiment. For each reflection, the area of q-space is kept constant with length of each side of 0.03 Å⁻¹. Following the evolution of each spot, the reference location is kept constant

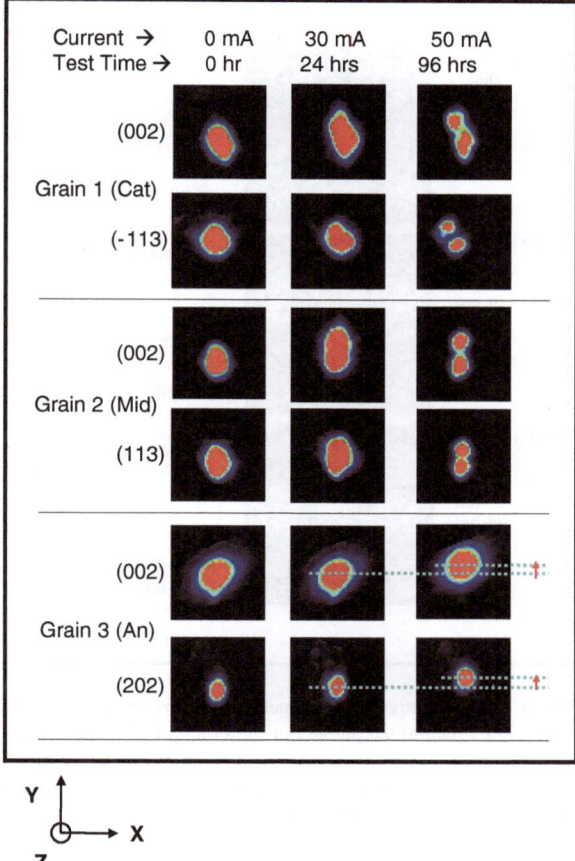

evident from Fig. 3.5a–c, that Grain 2 has evolved from initially, an undeformed crystal, to a plastically-bent crystal, and then lastly, to polygonized sub-grain structures, such as illustrated in Fig. 3.5d. Each of the features here illustrates the corresponding Laue diffraction observations on the left side. From the amount of broadening we can calculate the bending of the Cu crystal, and from the amount of splitting, the angle of misorientation.

We can then use the broadening and the spot splitting observed to obtain information about the dislocation structure in the grain induced by electromigration. For instance, from the streak length of Fig. 3.5b as measured in the CCD camera and the sample to detector distance we obtain the curvature angle of the grain of $0.75°$. Since the mapping of the out-of-plane orientation of the crystal along the Cu line indicates a near bamboo structure, the grain width is about the same as the width of the line (1.6 μm), from which we get the radius of curvature of the grain, R = 126 μm. The geometrically necessary dislocation density to account for the curvature observed can be calculated from the Cahn-Nye relationship [14–15] $\rho = 1/Rb$ where b is the Burgers

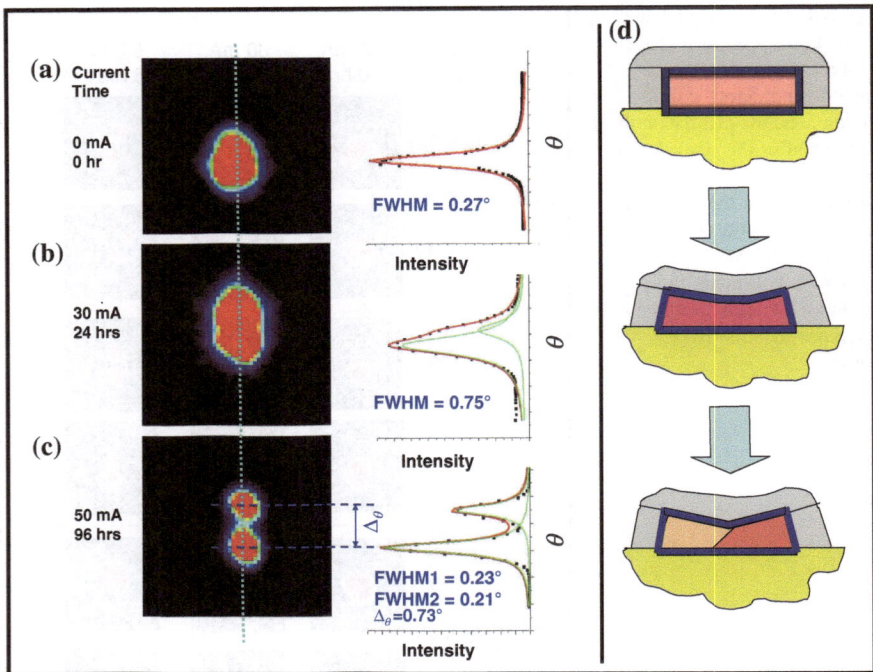

Fig. 3.5 Quantitative measurement of the evolution of a Laue diffraction spot of Grain 2 of the three grains mentioned above during an in situ EM experiment; Broadening is shown as the spot evolves from (**a**) to (**b**), while peak splitting, signifying the formation of low-angle grain boundaries, is evident in (**c**) as EM progresses; The evolution of Grain 2, in the cross-section of the line, is illustrated from undeformed, to plastic bending to formation of sub-grain structures in (**d**)

vector. The geometrically necessary dislocation density is then $\rho = 3 \times 10^9/cm^2$. The total number of dislocations introduced is only 49.

To obtain quantitative information on polygonization walls (small angle grain boundaries) from the spot split in Fig. 3.5c we observe that the Laue spot splitting, $\Delta_\theta = 0.73°$. From this misorientation and Burgers' model of a small angle grain boundary $\Delta_\theta = b/L$, where L = dislocation spacing, we find $L = 212$ Å which amounts to 45 dislocations in the boundary.

3.4.3 Electromigration-Induced Plasticity: The Linewidth Effects

We now describe in situ electromigration studies on another damascene Cu test structure with different linewidth (0.6 μm). The higher test temperature of this group of experiments was designed to give more pronounced streaking of the Laue peaks as the grains undergo electromigration. A previous similar electromigration

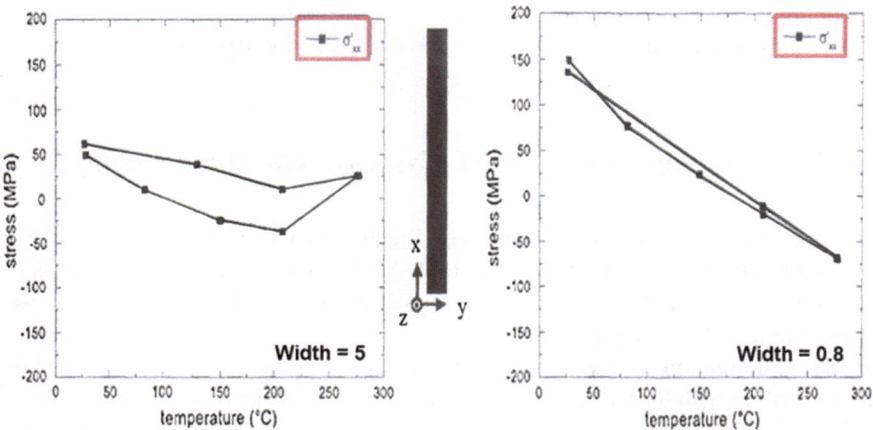

Fig. 3.6 Deviatoric stress (σ_{xx}') versus temperature curves suggest plastic deformation in wide Cu line, but not in the narrow geometry—consistent with our EM plasticity observation in Cu lines in the present study (courtesy of Spolenak et al. [16])

study has shown an extensive broadening of peaks in the Al(Cu) system [5]. By increasing the test temperature of this group of experiments, we aim to have a comparable homologous temperature (T/T_M) to that of the previous study on Al(Cu), which was ~ 0.51. The homologous test temperature (T/T_M) for the present study of narrow Cu lines was ~ 0.48, which is higher than homologous temperature for the wide line experiments discussed above (~ 0.4).

However, our observation of the peaks of the grains in the narrow Cu line (width = 0.6 μm) did not show any broadening of the peaks during electromigration. This is true despite the higher temperature used in this group of experiments. Instead, grain rotations, similar to that of Grain 3 in Fig. 3.5, are observed throughout the length of the line. The rotation of grains manifests itself as a shifting in the position of the Laue spot; from the direction and magnitude of the shift, we can calculate the axis and amount of the crystal rotation (Fig. 3.6).

The narrow Cu line thus appears to behave less plastically in respond to similar EM current density stressing, compared to the wide (1.6 μm) line (as has just been described in Sect. 3.4.2). Narrower Cu lines thus seem to have higher electromigration resistance (compared to wider Cu line). The resistance to plastic flow and the reasons why only grain rotation occurs in the narrow Cu line are not well understood at present. However higher resistance to plastic deformation in smaller structures, especially in the case of Cu line structures, has also been reported by Spolenak et al. [16].

In that study [16], thermal cycling of wide Cu damascene line (5 μm) was shown to exhibit plastic deformation with yield stresses in the range 50–100 MPA in compression. In contrast, the narrower line (0.8 μm) in that study indicated no such yield behavior and deformed only elastically over the entire temperature cycling range. These observations are consistent with our electromigration results.

Streaking of X-ray Laue spots indicating plastic bending of grains is observed for the wider 1.6 μm line, but is absent for the narrower line (0.6 μm).

3.4.4 Electromigration-Induced Plasticity: The Directionality

We discussed earlier about the consistent direction in which this electromigration-induced plasticity has been observed both in Al [5] interconnect lines, as well as Cu (this present study). Here, we further investigate the directionality of the plastic deformation in term of electromigration current direction. We use the Laue peak streaking simulation to study the possible mechanisms of plasticity. By matching the simulation with the actual streaking pattern, we can make some conjectures as to which particular slip system—among the 12 possible slip systems known for FCC metals—might be responsible for the streaking pattern.

Coming back to the wide lines, Fig. 3.7 shows the movement of Laue spots of Grain 2 during the EM test, and a comparison with simulation result. Working within the Cahn-Nye [14, 15] crystal bending model, and knowing the initial and end states of each of the diffraction spots, we can simulate the dislocation processes in the crystal necessary to cause the transformation of diffraction spots from their initial state to their final streaked/split/shifted state. We discovered that indeed the movement of the diffraction spots in Fig. 3.7a can be simulated by certain dislocation slip processes belonging to a slip system known to operate in FCC crystals (Fig. 3.7b). Figure 3.8a then describes the particular slip plane, slip direction and line direction in Grain 2 with respect to the interconnect line coordinates that are involved in the respective movement of diffraction spots.

Grain 2 is a large grain spanning across the cross-section of the line, as thus can be modeled as shown in Fig. 3.8b. The simulation shows one possible scenario of evolution of the grain as electromigration progresses. Point defect transport on the interfaces of the Cu line initiates the production of dislocations within the crystal of Grain 2 and activates their movement. Dislocations glide on a {111} plane that is tilted 41° from the surface of the sample. As dislocations accumulate on the glide planes, they become unstable with respect to climb and begin to coalesce into tilt dislocation walls as illustrated in Fig. 3.8c. This series of events would manifest itself in the form of split Laue peaks following a certain direction as experimentally observed during electromigration to Grain 2 as shown in Fig. 3.7a.

We also observe that the ⟨112⟩ type direction for the tilt axis of the crystal to be very close (within a few degrees) to the direction of the electron flow, or in other words, to the direction along the length of the line. The example in Fig. 3.8b shows a 6° deviation between the axis of tilt, which is a ⟨112⟩, and the direction of electron flow. More specifically, upon further inspection, this particular ⟨112⟩ was also found to be the closest ⟨112⟩ in Grain 2 to the direction of electron flow.

Fig. 3.7 (**a**) Laue reflection spots at the initial stage (*orange*) and after they split into the second set of reflection spots (*yellow*); (**b**) simulation of the same initial set of reflection spots streaked based on a particular slip system showing a match with experiment (**a**)

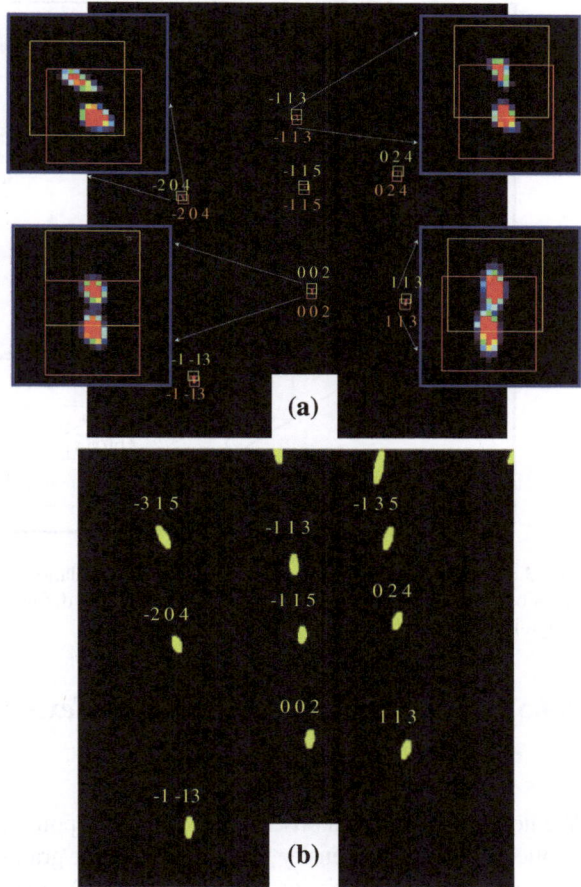

Further, more complete studies confirmed that this observation holds true for all the Cu grains in the current experiment exhibiting plastic bending and/or polygonization (formation of subgrain structures). This is shown in Fig. 3.9a.

Nine large Cu grains with the observed plastic bending/polygonization are represented by the 9 bullets within 10° of the closest ⟨112⟩ direction of their respective crystal to the direction of the electron flow. The bullets represent the direction of the in-plane crystallographic orientation of each of the respective grains in the direction of the electron flow in the line. To put it simply, all grains with a ⟨112⟩ direction pointing in the direction of the length of the line were observed with plastic bending/polygonization. More specifically, it has also been consistently observed that the particular ⟨112⟩ direction then becomes the tilt axis of the plastic deformation in the respective grain. This suggests a correlation of the proximity of certain ⟨112⟩ line directions to the direction of electron flow with the occurrence of plastic behavior.

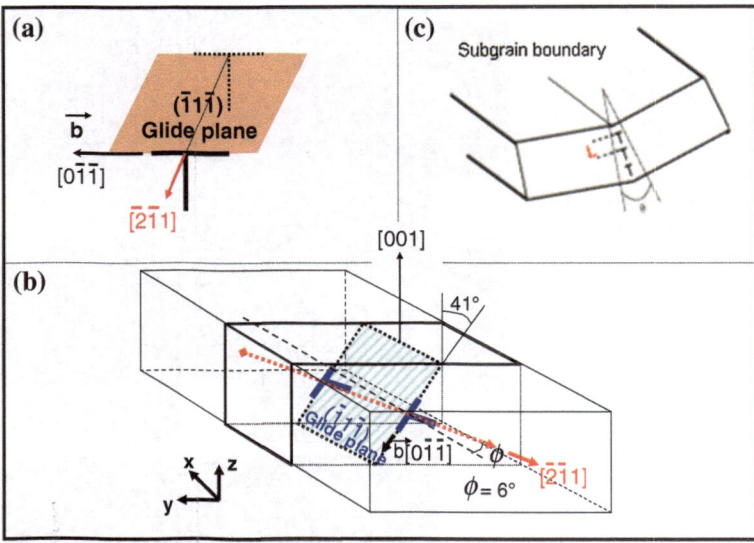

Fig. 3.8 (**a**) The active slip system for which the simulation predicts the movement of reflection spots correctly in Fig. 3.7a and b; (**b**) modeling of slip deformation in Grain 2; (**c**) illustration of subgrain boundary formation through polygonization

3.4.5 Correlation Between In-Plane Texture and Occurrence of Plasticity

We now recap our main observations up to this point. Almost immediately upon the application of high-density current flow, Cu grains behave plastically (crystal bending, polygonization, or simply rotation). We confirm that when large single grains spanning across the cross-section of the Cu lines are exhibiting crystal bending (and/or crystal polygonization), the rotation axis of this plastic deformation —which is always one of the $\langle 112 \rangle$ of the 12 possible slip systems in FCC crystals —has been found to be very close (within 10°) to the direction of the electron flow, as illustrated in Fig. 3.9a. This particular $\langle 112 \rangle$ is also always the closest $\langle 112 \rangle$ to the direction of electron flow. This finding thus suggests that the proximity of a $\langle 112 \rangle$ direction in the grain to the direction of the length of the line is correlated to the occurrence of plastic behavior in a given grain.

This particular finding might have an important practical implication. If all the grains exhibiting crystal bending/polygonization are those with a $\langle 112 \rangle$ direction that falls closely within the direction of the electron flow, then we could propose that if a grain is oriented such that none of its $\langle 112 \rangle$ direction falls within 10° of the electron flow direction, then this grain would be less susceptible to plastic deformation induced by electromigration. Such grains would have their crystallographic orientations in the direction of the electron flow in the lines fall within the blue-colored regions as shown in Fig. 3.9b. As such, it could be proposed that these are

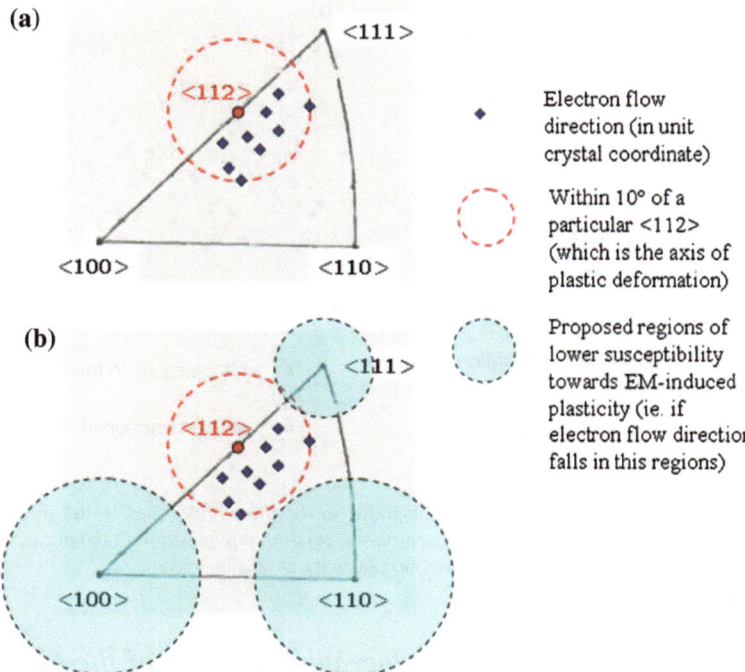

Fig. 3.9 Proximity between a particular ⟨112⟩ direction (which is also the axis of plastic deformation) with the direction of the electron flow in the crystal is suggested to have correlation with the occurrence of EM-induced plastic deformation (**a**) nine large Cu grains with the observed plastic bending/polygonization are represented by the 9 bullets within 10° of a particular ⟨112⟩ direction; (**b**) the *blue regions* represent preferred in-plane textures proposed to give higher resistance towards EM-induced plasticity

the particular in-plane textures of Cu interconnect lines that might give the wires lower susceptibility towards EM-induced plastic deformation, or in other words, higher resistance to EM later damage.

Finally, a full, more complete correlation study involving those other grains not exhibiting plastic bending/polygonization in this experiment has been conducted to confirm this hypothesis. The results are shown in Fig. 3.10. Crystallography of each of the grains in these wide Cu lines was examined in correlation with the direction of the electron flow in the lines. We show here in Fig. 3.10a evidently that all grains not exhibiting plastic bending and/or polygonization (red triangles) in this experiment, except for two data points, are largely pointing either ⟨100⟩ or ⟨110⟩ in the direction of the length of the line—i.e. not ⟨112⟩, or anything close to it. Upon further examination, we show in Fig. 3.10b that five data points among the red triangles belong to grains in which plastic deformation actually happens, but only in the form of crystal rotation. This group includes the two red triangles within 10° from ⟨112⟩ direction. This suggests that while the trend is clear for plastic bending and/or polygonization, grain rotation results show mixed correlation.

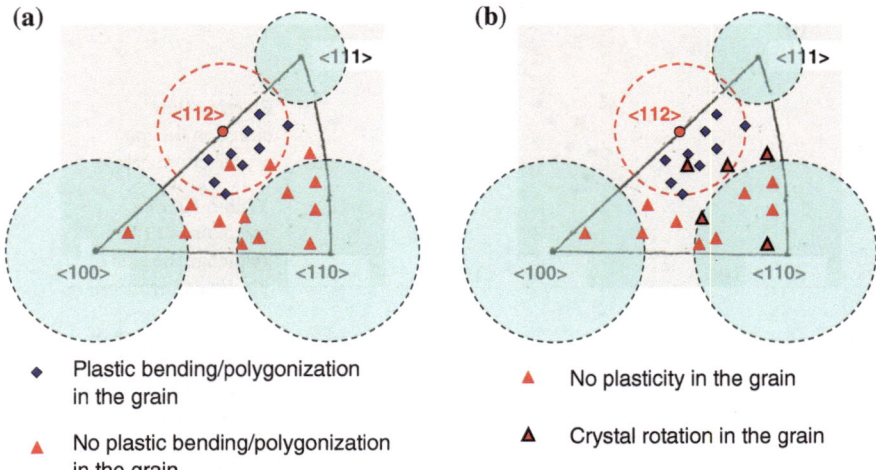

Fig. 3.10 Full correlation involving all Cu grains in the three wide lines in the present study (a) plastic bending/polygonization versus non-plastic bending/polygonization; (b) plastic bending/polygonization versus grain rotation versus no plasticity at all observed

3.4.6 The Out-of-Plane Crystallographic Texture of the Cu Lines

Mapping of out-of-plane orientations of the Cu grains in the wide lines using the technique of white-beam X-ray microdiffraction was conducted and discussed in Fig. 3.3 (Sect. 3.4.1). We further aimed for quantitative analysis to determine texture of the wide Cu lines in the present study. This was done by examining crystallographic orientations of all the grains in the Cu lines, and counting and grouping them. The results are shown in Fig. 3.11. This is the texture result of a particular wide Cu line, but is typical also of the other two wide Cu lines investigated in the present study.

Figure 3.10 shows widely varying out-of-plane orientations of the grains in this line. This observation is in contrast to the texture known to form in sputtered Cu films. However, it is in agreement with observations made on annealed electroplated films [17]. The (115) twin texture has been found to be predominant. This is verified in the present experimental result. At least a third of the grains have their $\langle 111 \rangle$ directions $38.9°$ off the normal of the line—the theoretical angle between a (111) plane and a (115) plane. Additionally, there are also $\langle 111 \rangle$ components and indications for a sidewall texture (grains having their $\langle 111 \rangle$ directions between $15°$ and $25°$ off the normal of the line), as shown in Fig. 3.11. This has been known as a feature of electroplated Cu damascene structures and has been reported previously. [11, 13] However, it is evident from Fig. 3.11 that there are also a significant number of many other (i.e. non-$\langle 111 \rangle$-texture-related) grains—almost as many—as

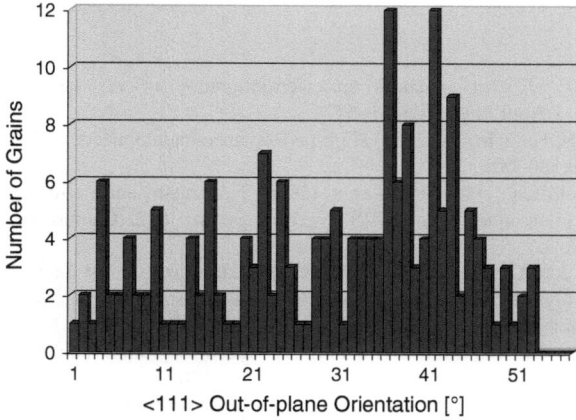

Fig. 3.11 Quantitative determination of texture in the Cu lines; number of grains found in the line versus ⟨111⟩ out-of-plane orientation (defined as the deviation between the ⟨111⟩ orientation of the crystal with the normal direction of the sample/line surface; the closer this numbers are to 0, the the closer the grains are to being ⟨111⟩-oriented)

there are ⟨111⟩ and their related twin and sidewall-growth orientations typical of electrodeposited Cu damascene lines. [14].

3.5 Conclusions

Using a synchrotron technique involving white-beam X-ray microdiffraction, we have observed plastic deformation behavior of Cu polycrystals during electromigration experiments. The extent of the plastic behavior was found to be dependent on width of the line test structure. Of all the grains in the line test structure, the few grains that exhibit bending and polygonization or rotation, tend to be the largest grains, spanning across the cross-section of the line, and having orientations with a ⟨112⟩ direction nearly parallel to the line corresponding to the activation of a known FCC slip system by the application of current. This finding suggests a correlation of the proximity of a ⟨112⟩ line direction to the direction of electron flow with the occurrence of plastic behavior, which may then lead to a practical, important implication in determining Cu texture of higher resistance to electromigration-induced plastic damage. We believe that crystal plasticity in small scale devices under some electrical loading may have a direct bearing on the performance and reliability of today's nanoscale devices.

References

1. Tu KN (2003) Recent advances on electromigration in very-large-scale-integration of interconnects. J Appl Phys 94:5451–5473
2. Havemann RH, Hutchby JA (2001) High-performance interconnects: an integration overview. Proc IEEE 89:586–601
3. Kuan TS, Inoki CK, Oehrlein GS et al (2000) Fabrication and performance limits of sub-0.1 μm Cu Interconnects. In: MRS Proceedings vol 612. Cambridge University Press, Cambridge, pp D7
4. Vanasupa L, Joo YC, Besser PR et al (1999) Texture analysis of damascene-fabricated Cu lines by x-ray diffraction and electron backscatter diffraction and its impact on electromigration performance. J Appl Phys 85:2583–2590
5. Valek BC, Bravman JC, Tamura N et al (2002) Electromigration-induced plastic deformation in passivated metal lines. Appl Phys Lett 81:4168–4170
6. Tamura N, MacDowell AA, Spolenak R et al (2003) Scanning X-ray microdiffraction with submicrometer white beam for strain/stress and orientation mapping in thin films. J Sync Rad 10:137–143
7. Meyer MA, Hermann M, Langer E et al (2002) In situ SEM observation of electromigration phenomena in fully embedded copper interconnect structures. Microelectron Eng 64:375–382
8. Vairagar AV, Mhaisalkar SG, Krishnamoorthy A et al (2004) In situ observation of electromigration-induced void migration in dual-damascene Cu interconnect structures. Appl Phys Lett 85:2502–2504
9. Doan JC, Lee S, Lee SH et al (2000) A high-voltage scanning electron microscopy system for in situ electromigration testing. Rev Sci Instr 71:2848–2854
10. Lingk C, Gross ME, Brown WL, Siegrist T, Coleman E, Lai WY-C, J. Miner F, Ritzdorf T, Turner J, Gibbons J, Klawuhn E, Wu G, Zhang F (1999) Advanced metallization conference 1998 (AMC 1998). Materials Research Society, Warrendale, PA, p 73
11. Lingk C, Gross ME, Brown WL (1999) X-ray diffraction pole figure evidence for (111) sidewall texture of electroplated Cu in submicron damascene trenches. Appl Phys Lett 74:682–684
12. Lingk C, Gross ME, Brown WL (2000) Texture development of blanket electroplated copper films. J Appl Phys 87:2232–2236
13. Besser P, Zschech E, Blum W et al (2001) Microstructural characterization of inlaid copper interconnect lines. J Elec Matls 30:320–330
14. Cahn RW (1949) Recrystallization of single crystals after plastic bending. J Inst Metals 76:121
15. Nye JF (1953) Some geometrical relations in dislocated crystals. Acta Metall 1:153–162
16. Spolenak R, Tamura N, Valek B et al (2002) Stress induced-phenomena in metallization. In: Baker SP, Korhonen MA, Arzt E et al (eds) American institute of physics conference proceedings, vol 612. AIP Publishing, Melville
17. Spolenak R, Volkert CA, Takahashi KM et al (1999) Mechanical properties of electroplated copper thin films. Mat Res Soc Proc 594:55

Chapter 4
Electromigration-Induced Plasticity in Cu Interconnects: The Texture Dependence

Abstract The discussion of the peculiarities of electromigration in Cu interconnects is continued in this chapter. It is shown that the interconnect texture is an important factor, governing the plastic response of Cu grains to the electrical current. The degree of the plastic response is proportional with the availability of $\langle 112 \rangle$ direction in Cu crystals along the direction of the current. Hence, (111) out-of-plane orientation of Cu grains increases the plastic effect of electromigration, while the presence of the grains with a different orientation could weaken it.

Keywords Elasticity · μXRD · Texture · Interconnect · Electromigration stress · Dielectric

4.1 Introduction

Most interconnect metals are aggregates of crystalline grains. The crystalline lattice of each grain has a characteristic orientation, and a polycrystal is thus characterized by a distribution of orientations—its texture. Texture governs many of the physical, electrical and mechanical properties of polycrystalline materials. In metallic conductor lines in microelectronics integrated circuits, texture has been known to play important roles in the performance and reliability of the conductors, for instance in electromigration [1].

Plastic deformation has been observed in damascene Cu interconnect test structures during an in situ electromigration experiment and before the onset of visible microstructural damage (i.e. voiding) using a synchrotron technique of white beam X-ray microdiffraction, as has been described and discussed in the previous chapter. We show here, in this chapter, that the extent of this electromigration-induced plasticity is dependent on the texture of the Cu grains in the line. In lines with strong $\langle 111 \rangle$ textures, the extent of plastic deformation is found to be relatively large compared to our plasticity results in Chap. 3 using another set of Cu lines with weaker textures. This is consistent with our earlier observation (detailed in Chap. 3)

© The Author(s) 2015
A.S. Budiman, *Probing Crystal Plasticity at the Nanoscales*,
SpringerBriefs in Applied Sciences and Technology,
DOI 10.1007/978-981-287-335-4_4

that the occurrence of plastic deformation in a given grain can be strongly correlated with the availability of a $\langle 112 \rangle$ direction of the crystal in the proximity of the direction of the electron flow in the line (within an angle of 10°). In $\langle 111 \rangle$ out-of-plane oriented grains in a damascene interconnect scheme, the crystal plane facing the sidewall tends to be a {110} plane, so as to minimize interfacial energy. Therefore, it is deterministic rather than probabilistic that the $\langle 111 \rangle$ grains will have a $\langle 112 \rangle$ direction nearly parallel to the direction of electron flow.

However, as we establish this texture-plasticity correlation, the effect of plasticity on the Cu EM degradation process remains ambiguous. We have models and preliminary observations suggesting that the effect of plasticity can be both beneficial as well as damaging to EM, perhaps even at the same time. Depending upon the effect (or net effect) of plasticity on EM thus, this chapter may offer control to the desirable texture of Cu interconnect lines in electromigration.

4.2 Background

4.2.1 Electromigration-Induced Plasticity in Metallic Interconnects

With the observation described and discussed in Chap. 3 of this book, as well as reported in the literature [2], we extend the generality of electromigration-induced plasticity phenomenon from aluminum interconnects [3] to copper interconnects as well. This phenomenon of plastic deformation in metallic interconnects was, at first, deemed unexpected [4–6] as the classical view of stress evolution during electromigration (EM) in metallic interconnects had hitherto involved only the hydrostatic stresses (hydrostatic tension build-up in the cathode end, and compression in the anode end, due to transport of point defects).

Not only did we observe plastic deformation in both Al and Cu lines during electromigration, we also consistently observed the same directionality of the plastic deformation in both metallic lines [2–6]. The grains that exhibit plastic deformation were always observed to deform (bend, rotate or polygonize) transverse to the direction of the electron flow in the line, or in other words, across the width of the line. This deformation geometry leads us to conclude that dislocations introduced by plastic flow lie predominantly in the direction of electron flow and may provide additional easy paths for the transport of point defects. Since these findings occur long before any observable voids or hillocks are formed, they may have direct bearing on the final failure stages of electromigration.

Linewidth and grain size effects were observed in Cu interconnects, and discussed in Chap. 3. In wide lines, plastic deformation manifests itself as grain bending and the formation of subgrain structures, while only grain rotation is observed in the narrower lines. Larger Cu grains spanning across the width of the lines tend to exhibit larger extent of plasticity (crystal bending or polygonization), while smaller grains rotates; this was also found to be true in the Al interconnects [3]. We further observed that the occurrence of plasticity can be correlated with the availability of a $\langle 112 \rangle$ direction in the proximity of the direction of the length of the line (within $10°$). This then became a basis of our argument proposed in this chapter; it also has important practical implications for the device reliability community in industry as will be further elaborated in Chap. 5.

In this chapter, we study a different set of Cu lines fabricated by a different manufacturer. This set of Cu lines differs with the previous set (as used in Chap. 3, as well as in literature [2]) in a few ways; chief among them is texture. Again using the synchrotron technique of white beam X-ray microdiffraction, we follow the evolution of plasticity in Cu polycrystals during similar electromigration experiments as in Chap. 3. We find strong texture dependence and propose a model to explain such prominent observation (using the correlation we observed from Chap. 3). In this set of samples, the Cu lines were surrounded by two different sets of dielectric materials. This has enabled us to also study the effect of dielectric constraints.

4.2.2 Microstructural Characterization of Cu Lines Manufactured by AMD

Inlaid Cu lines fabricated by different manufacturers have been known to have different crystallographic textures [7, 8]. As was the experience with Al interconnects, the fabrication method strongly influences grain growth [9–13] and mechanical stress [14, 15]. Inlaid Cu processing [16–18] is fundamentally different from the subtractive etch process of Al, in which the microstructure of the line is controlled by Al grain growth in the blanket film (thin film). The Cu is typically electroplated onto a Cu seed and thus the microstructure originates from the trench. The microstructure in inlaid Cu interconnects thus depends on many more parameters, a few of the most critical of them are process-related parameters, such as the conditions of the electroplating process (chemistry, additives, waveform, etc.), the barrier materials used, or the Cu seed layer thickness) [7].

As we intend, in this chapter, to study the evolution of crystal plasticity/plastic deformation in Cu interconnects manufactured by AMD, it would be of paramount importance to understand the typical microstructures observed in these inlaid Cu lines. Besser et al. [7] in his 2001 study, has characterized in great detail the grain size, crystallographic texture and mechanical stresses of Cu lines of various widths made also by AMD, and it is useful to summarize the results here.

The microstructure of inlaid Cu lines was quantified as a function of annealing conditions, post-plating and post-chemical-mechanical-planarization (CMP). The grain size distribution was measured using the median intercept method, crystallographic texture was characterized by pole figure analysis, and mechanical stress was determined using X-ray diffraction. The median grain size and mechanical stress level was found to increase with increasing annealing temperature. The

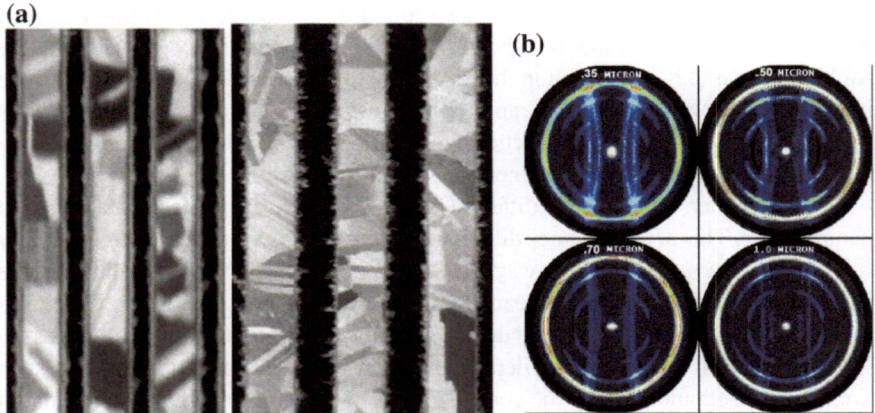

Fig. 4.1 Strong (111) texture observed in inlaid Cu interconnects [6]; **a** FIB images of 0.35 μm (*left*) and 1.06 μm (*right*) lines; **b** Crystallographic texture results as a function of linewidth for inlaid Cu lines; the fraction of sidewall-nucleated (111) grains increases with decreasing linewidth (courtesy of Besser et al. [7])

crystallographic texture is independent of the anneal temperature and is predominantly (111) with a small fraction of sidewall-nucleated (111) grains, such as shown in Fig. 4.1 (courtesy of Besser et al. [7]).

As the linewidth is decreased from 1.06 to 0.35 μm, the (111) intensity in the pole figure reduces, signifying the reduced contribution of trench bottom-nucleated (111) grains and simultaneously the increasing portion of the sidewall-nucleated (111) grains in the inlaid Cu lines [7]. This grain growth in the trench is independent of that in the overburden.

It is also important here, for reasons that will be shown later in this chapter, to note that the (111) grains nucleated from the trench bottom have a preferred in-plane orientation [7], such as shown in Fig. 4.2a—(courtesy of Besser et al. [7]). This has been supported by other studies [10, 19, 20]. Considering the three-fold symmetry of the (111) plane, the preferred in-plane orientation to the (111) can only arise from its preference to have a [aa0] sidewall, such as illustrated in Fig. 4.2b (again courtesy of Besser et al. [7]). This is a preferred orientation since it will minimize the interfacial energy of the grains [19].

4.2.3 Influence of Dielectrics on Mechanical Stresses and Plastic Deformation

Mechanical stresses in interconnect lines mostly arise from the difference in the coefficient of thermal expansion (CTE) between the interconnect metal and the substrate/dielectric that constraints it. As was the experience with Al, the fabrication method is a strong influence in the mechanical stress state of Cu interconnects.

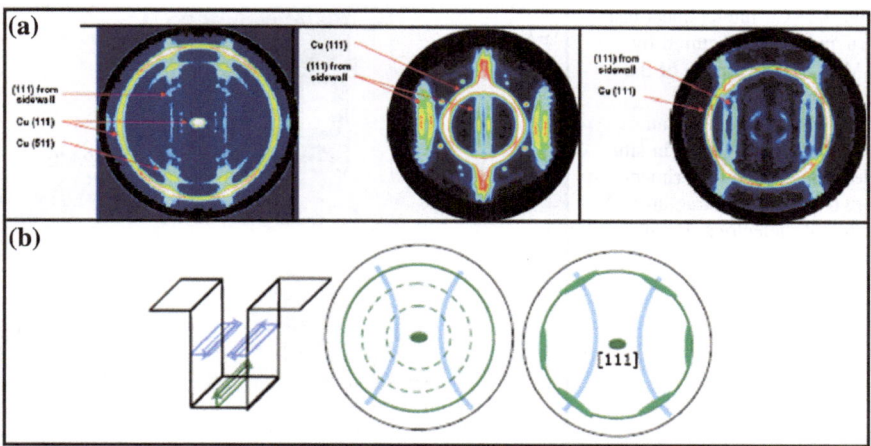

Fig. 4.2 Preferred in-plane orientation to the (111) texture observed in inlaid Cu interconnects [7];
a The (111), (110) and (100) pole plots for 0.35 μm lines (*left, middle,* and *right* in **a**); the (111)
grains prefer to have a [aa0] sidewall; **b** Schematic drawings illustrating the development of (111)
texture in lines and the influence of these texture components on the pole plot (*left* and *middle*
drawings); the drawing on the *right* shows the pole plot for line with vertical sidewalls and a
preferred [aa0] in-plane (111) orientation (courtesy of Besser et al. [7])

Inlaid processing [18, 21] is fundamentally different from the subtractive etch
process, and its effect on the stress state has been documented [14, 15]. Large
thermal stresses may build up during the successive thermal cycling, due to the
differences in the coefficient of thermal expansion (CTE) of the component materials [22]. Furthermore, the adoption of various low-k materials as dielectrics has
aggravated the situation.

Since the magnitude and state of the stresses in Cu interconnect lines are greatly
affected by the thermomechanical properties of the surrounding dielectric materials,
this in turn is expected to result in substantially different failure behaviors. For
example, using a combination of synchrotron X-ray diffraction and finite element
analysis (FEA), Paik et al. [23] found that the von Mises stress of Cu/low-k is larger
than that of Cu/SiO$_2$ due to the high CTE of low-k materials (SiLK, for instance,
has a CTE of 62.0 as cited in Ref. [23]). Therefore the Cu in Cu/low-k scheme is
vulnerable to extreme plastic deformation. Already, via failures have been reported
[24, 25] due to deformation by shear stresses, rather than by hydrostatic stresses.
This has also been supported by Shen et al. [26] using 3D numerical modeling/
analysis.

This mode of failure due to plastic deformation is thus likely to continue to be
one of the main failure modes in the next-generation interconnects, as long as they
are still made with low-k materials. Therefore, the understanding of plastic deformation and especially the role of dielectric constraints could be the key to
improving reliability of today's and future Cu/low-k interconnects.

Fig. 4.3 Cu interconnect test structures manufactured by AMD; **a** SEM image of the test structure (*Acknowledgement* Bryan Tracy of Spansion); **b** in situ electromigration experiment; **c** two sets of test structures of different dielectrics: low-k versus hybrid

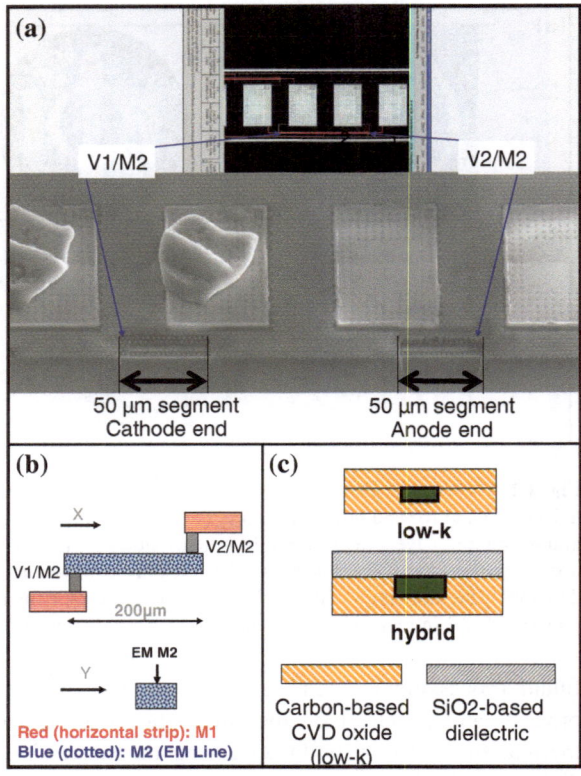

4.3 Experimental

The interconnect test structure used in this study is an electroplated Cu damascene line manufactured by AMD (Fig. 4.3), not unlike the inlaid Cu lines used by Besser et al. [7] (they come from the same fabrication methodology). The M2 lines were studied after partial removal of the top dielectric to expose the capped Cu lines (Fig. 4.3a). In this technology, the Cu fill process includes a standard Ta-based barrier and Cu seed, electroplated Cu fill, post-plating anneal, chemical-mechanical polish and a dielectric cap layer. Two different inter layer dielectrics (ILD) were integrated with copper: Cu/low-k ILD (Low-k = CVD carbon-doped oxide) and Cu/Hybrid ILD (Hybrid = Cu/low-k at the line level and Cu/FTEOS at the via layer). Both ILD materials were studied in order to provide a comparison of the extent of plasticity.

The first set of test structures consists of lines 200 μm length, approximately 0.2 μm thick, and 0.5 μm wide. Due to limited beam time, typical of synchrotron experiments, only segments of 50 μm length at both cathode and anode ends of the line were studied (Fig. 4.3a). The dielectric is carbon-based CVD oxide ("low-k" in Fig. 4.3c). The second set of interconnect test structures was prepared with dimensions similar to those of the first one, but with the hybrid ILD material (SiO2 based). The line length is 200 μm, the thickness is approximately 0.25 μm and the width is 0.7 μm. Similarly, only segments of 50 μm were studied.

The synchrotron technique of scanning white beam X-ray microdiffraction has been described in a complete manner in a previous chapter of this book, as well as in the literature [27]. It consists of scanning the sample under a submicron size X-ray beam and capturing a Laue diffraction pattern at each step with a CCD detector. Using a submicron-sized beam allows us to consider each grain of the interconnect sample as a single crystal. The indexing of the Laue pattern gives the orientation of the grain while the shape of the Laue peaks yields information regarding plastic deformation of the individual grain.

The X-ray microdiffraction experiment was performed on beamline 12.3.2. at the Advanced Light Source, Berkeley, CA. The electromigration test was conducted at 300 °C on a via-terminated test structure (Fig. 4.3b). Current and voltage were monitored at 10 s increments. The sample was scanned in 0.5 μm steps. A complete set of CCD frames takes about 6–7 h to collect. The exposure time was 20 s plus about 10 s of electronic readout time for each frame. In this manner the Laue pattern and information regarding plastic deformation for each grain in the sample was collected for each time step during the experiment. The current was ramped up to 2 mA ($j = 2$ MA/cm^2) and then set at that value for the rest of the test (up to 36 h).

4.4 Results and Discussions

4.4.1 EM-Induced Plasticity: Directionality and Extent

We first describe the in situ electromigration observations on both of the Cu damascene test structures. Figure 4.4a–c show the typical evolution of the Laue diffraction spots during the in situ EM test. Figure 4.4 is early in the EM test, sampled after 36 h of testing. The observed broadening of the Laue diffraction spots (streaking) represents plastic deformation of the Cu grains induced by EM.

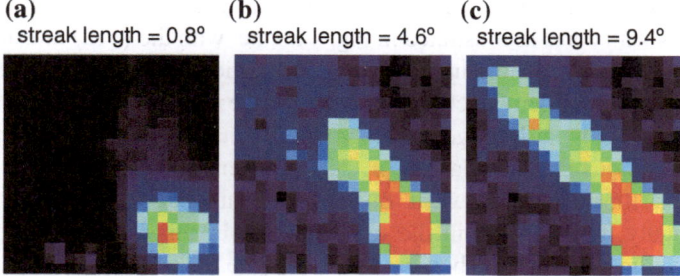

(a) streak length = 0.8° **(b)** streak length = 4.6° **(c)** streak length = 9.4°

Fig. 4.4 The typical evolution of Laue diffractions spots from Cu interconnect test structures during in situ electromigration experiments. **a** t = 0 h; RT, $J = 0$. **b** t = 20 h; T = 300 °C, $J = 2$ MA/cm^2. **c** t = 36 h; T = 300 °C, $J = 2$ MA/cm^2

Fig. 4.5 **a** Streaking and/or splitting of Cu Laue diffractions spots throughout a segment of the line; **b** Dislocations were found with cores lining up with the direction of the electron flow in the line (consistent with earlier observation discussed in Sect. 3.4) across grains throughout the length of the segment of the line observed (*asterisk* Grain map is estimated based on streaking observation)

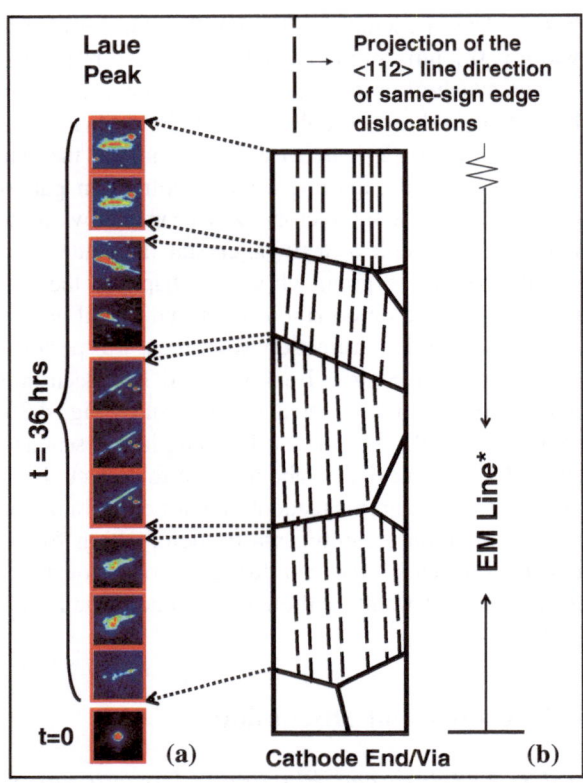

We find that as the EM test progresses, plasticity is observed in the Cu grains throughout the line, such as demonstrated in Fig. 4.5a. Not only is plasticity observed, but also that the direction of the plastic deformation is generally consistent across grains throughout the segments of the Cu interconnect line under observation, as shown in Fig. 4.5a. This is consistent with our observation on the previous set of Cu lines discussed in Chap. 3 (Sect. 3.4.5). Cu grains plastically deform in a direction transverse to the electron flow direction in the line. Such directionality can simply be accommodated by a distribution of same sign edge dislocations with cores as illustrated in Fig. 4.5b, i.e. with the $\langle 112 \rangle$ line direction of the dislocations are all lining up nearly to the direction of electron flow in the line.

The extent of plasticity such as indicated by the Laue spot streaking shown in Figs. 4.4 and 4.5 is fairly large, and represents a marked departure from our previous observations on a similar set of Cu damascene lines discussed in Chap. 3. Figure 4.6a illustrates the extent of the large amount of plasticity observed experimentally in Cu grains in the test structure. In this Laue image of the Cu interconnect line with the Carbon-based CVD oxide dielectric (after 36 h of EM at 300 °C and 2 MA/cm^2) most diffraction spots are shown to be broadened or split to various extents. The one set of diffraction spots that are relatively sharp and rounded belong to the Silicon substrate crystal. The most extreme broadening is

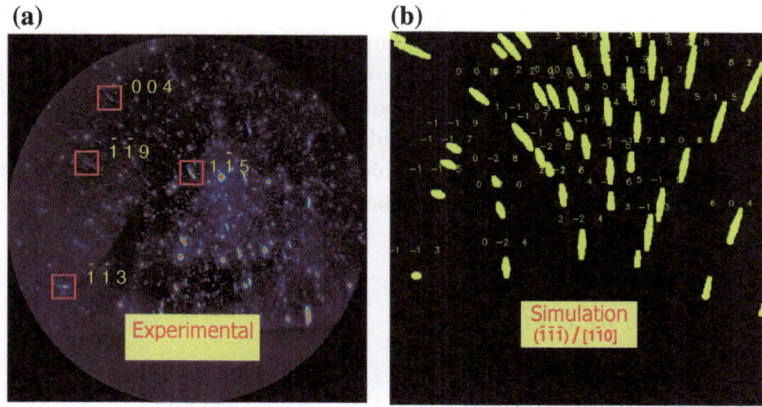

Fig. 4.6 **a** The general plasticity observed experimentally in Cu interconnects; **b** The streaking simulation of the Laue pattern of a particular Cu grain represented by a set of Laue diffraction spots (boxed in **a**)

shown by a set of Cu diffraction spots marked by boxes, which represents a Cu grain undergoing intense EM-induced plasticity.

The nature of plastic deformation exhibited by this particular grain can be further investigated by comparing the initial crystal orientation of the grain (through indexation of the diffraction spots), and a set of possible simulated streaking directions that represent the 12 possible plastic bending events in the FCC crystal. If one of the simulated streaking directions match with the actual streaking directions observed in Fig. 4.6a, we could infer the active slip system of the plastic deformation. The out-of-plane orientation of this particular Cu grain is close to $\langle 111 \rangle$, which is consistent with the typical texture of Cu lines seen in the industry [7]. In this approximately $\langle 111 \rangle$ oriented grain, the simulation (Fig. 4.6b) suggests that the $(\bar{1}\bar{1}\bar{1})/[1\bar{1}0]$ slip system is active, causing the crystal to deform plastically and to lead to the pattern of the diffraction spot streaking. The specific $\langle 112 \rangle$ line direction of this plastic deformation, which essentially becomes the rotation axis of the crystal bending, is observed to be $[-1-112]$ and that it is very close to the direction of the electron flow (off by 7.9°) in the interconnect line. This observation holds true (within 10°) throughout the many grains along the Cu line, and again is consistent with our previous observation in Chap. 3 (Sect. 3.4.5). This further confirms that the observed plasticity leads to the concentration of edge dislocations in Cu grains with cores running along the direction of electron flow throughout the full length of the interconnect lines such as illustrated earlier in Fig. 4.5b.

Exact grain orientation mapping of these Cu lines unfortunately could not be obtained in the present study. The X-ray spot size (0.8 × 0.8 μm) used in ALS beamline 12.3.2 was relatively large for the dimensions of these state-of-the-art interconnect lines. That makes diffraction spot indexation often very difficult and thus mapping of grain orientations and other further quantitative analyses unreliable. The one Cu grain that we discussed and analyzed above was among the limited number of grains in the two Cu lines for which indexation of the diffraction spots happens to be sufficiently clear and unambiguous for this analysis. In general, the larger the Cu grains, and the more bamboo-like they are, the more they diffract sharply and give numerous diffraction spots, thus giving higher confidence on the reliability of these results. That being said, however, it is fortunate that we could still always compare qualitatively the evolution of Cu diffraction spots before and after some period of EM testing, as has been demonstrated in Fig. 4.4.

4.4.2 Influence of Dielectrics

We now discuss the influence of dielectrics on our EM-induced plasticity observation. Figure 4.7a, b show still different additional diffraction spots observed during this experiment (after EM testing of 36 h, at 300 and 2 MA/cm^2 current loading) coming from the Cu lines with the low-k and the hybrid dielectrics, respectively. The diffraction spots have been converted to χ-θ space, with χ running along the direction of the length of the line, and θ across the direction of the width of the line. We can then use the broadening and the spot splitting observed to obtain information about the dislocation structure induced into the grain by electromigration.

For instance, from the streak length of Fig. 4.7a, as measured in the CCD camera and the sample to detector distance we obtain the curvature angle of the grain of 9.8°. Assuming a near bamboo structure, the grain width is the same as the width of the line (0.5 μm), from which we get the radius of curvature of the grain, $R = 2.34$ μm. The geometrically necessary dislocation density to account for the curvature observed can be calculated from the Cahn-Nye relationship $\rho = 1/Rb$ where b is the Burgers vector. The geometrically necessary dislocation density is then $\rho = 1.68 \times 10^{15}/\text{m}^2$. The total number of dislocations in the area of the cross-section of the Cu line/grain is approximately 142.

To obtain quantitative information on polygonization walls (subgrain boundaries) from the spot split in Fig. 4.7b we observe that the Laue spot splitting,

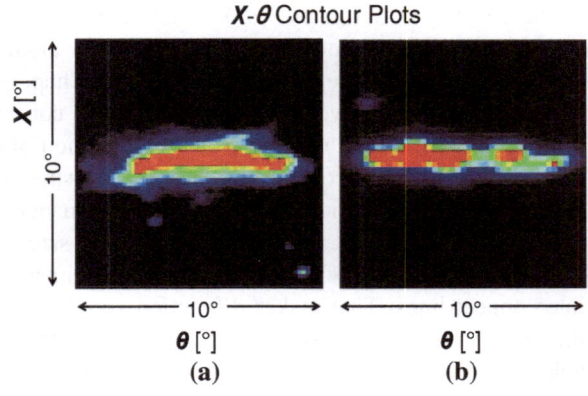

Fig. 4.7 The dielectric effects: the Laue peak streaking/splitting observed from Cu interconnect test structures with **a** low-k, and **b** hybrid dielectrics; in χ-θ space/contour intensity plot

$\Delta_\theta = 9.1°$. From this misorientation and Burgers' model of a small angle grain boundary $\Delta_\theta = b/L$, where L = dislocation spacing, we find $L = 16$ Å which amounts to a total of 110 dislocations in the subgrain boundaries in the cross-section of the Cu line/grain. This translates to the density of the geometrically necessary dislocation density of $\rho = 1.12 \times 10^{15}/m^2$.

The extent of the plasticity such as described here ($\rho \sim 10^{15}/m^2$) is observed across all grains throughout the segments of both the lines ("low-k" vs. "hybrid") that we studied. The significance of the difference in our analysis above, in terms of the extent of the plasticity, as well as, its manifestation (grain bending vs. polygonization) between the two Cu lines requires added confirmation. Nevertheless they provide a sense of the generality of the extent of plasticity in these Cu lines.

4.4.3 Proposed Correlation: Texture Versus EM-Induced Plasticity

Compared to the typical observation of the extent of the EM-induced plasticity in the previous set of Cu interconnect lines discussed in Chap. 3, this set of samples exhibit at least a two-order of magnitude difference, in term of GND density (Fig. 4.8). The former samples exhibit $\rho \sim 10^{12}$–$10^{13}/m^2$ (from here onwards, we call this "Samples A"), and the latter exhibits $\rho \sim 10^{15}/m^2$ ("Samples B"). As a reminder, Samples A and B were fabricated by different manufacturers and differ fairly significantly in dimensions, as well as dielectric materials used, such as illustrated in Fig. 4.8 (the dimensions are to scale).

Figure 4.8 shows the typical evolution of the Laue reflections from the Cu lines from initial state (RT, $j = 0$, t = time = 0) to EM state (after some electromigration, $T = 300\ °C$, $j \sim 2.0$–3.1 MA/cm^2, $t \sim 36$–96 h). Care was taken in order for the observed intensity contours in the χ-θ coordinate in Fig. 4.8 to be comparable (the windows all cover areas of squares of a range of 7°–10° in angle space, and the threshold of the lower-bound intensity display was set to be similar). Thus it is obvious from the relative apparent difference in the extent of streaking/splitting of the Laue diffraction spots that the level of plastic deformation that developed during the course of electromigration in Samples B is distinctively larger than that of Samples A. This has been further confirmed by our calculations of GND density earlier in this chapter (for Samples B), as well as in previous chapter (for Samples A) resulting in the typical ρ_{GND}'s displayed in Fig. 4.8.

As the two sets of samples (Samples A vs. B) under investigation are provided by different integrated circuit manufacturers, it is not possible to completely quantify the process differences (dielectric type, materials processing and thermal history) in their technologies in this book. It is known that the two sets of samples differ in terms of dimension and dielectric materials used; however, it is assumed that the main difference, as far as EM-induced plasticity is concerned, is the crystallographic texture of the Cu lines. From the texture analysis discussed in Chap. 3 (Sect. 3.4.6), we learn that Samples A have a rather weak (111) texture. Meanwhile, Samples B came from the same manufacturer of the inlaid Cu lines

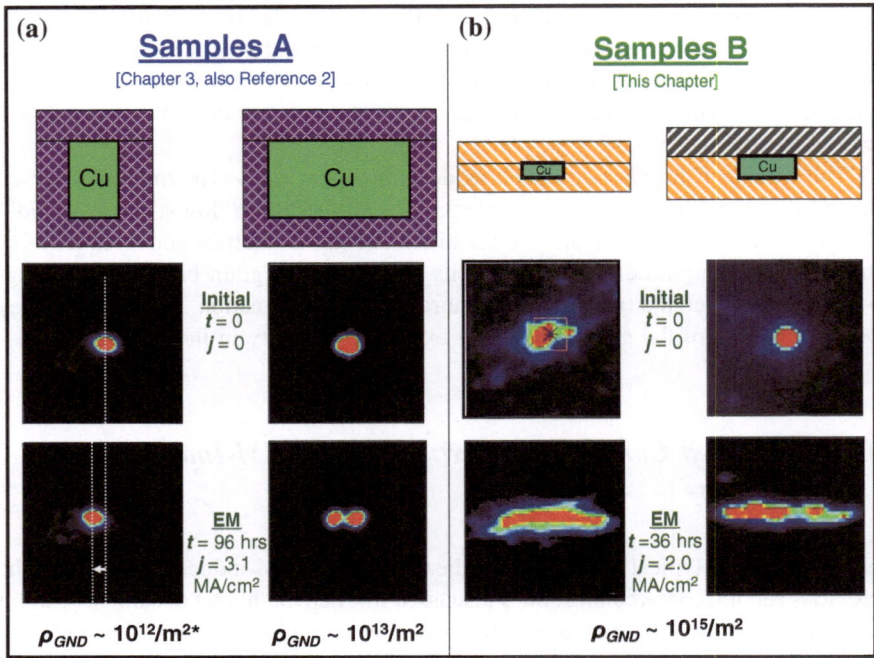

Fig. 4.8 Samples A versus B comparison: the schematic of the cross-sections (color code refers to different materials), typical evolution of Cu Laue diffraction spots (from "Initial" to "EM" states), and lastly, the typical densities of GNDs implied by the extent of streaking/splitting of Laue peaks; **a** Samples A (Cu lines reported in Chap. 3, as well as in the Ref. [2]); **b** Samples B (Cu lines reported in this chapter); They both were fabricated by different manufacturers. (*asterisk* taken as that of typical annealed metals)

studied by Besser et al. [7]. It is therefore reasonable to assert that Samples B would have the typical strong (111) texture observed by Besser et al. [7]. Unfortunately, a grain orientation mapping using Beamline 12.3.2 in the present study was not possible for Samples B for reasons that have been discussed in the earlier part of this chapter (Sect. 4.4.1).

While other process and dimension differences between these two sets of samples include dimensions and dielectric materials are acknowledged, we believe that these differences cannot satisfactorily explain the differences in the extent of plastic deformation. For example, the Cu in Samples A is surrounded completely by dielectric material, which is a fluorinated SiO_2-based dielectric, and thus generally believed to constrain the Cu lines better and should result in less plastic deformation. This is consistent with our observation of Samples A versus Samples B, but the different dielectric schemes in Samples B, do not appear to affect the level of plasticity in the Cu lines (discussed in Sect. 4.4.2). Another example involves the size effect. Wider lines seem to exhibit more plastic deformation as discussed in Chap. 3, such as also shown in Fig. 4.8a. However, Samples B actually are much narrower, and also much smaller in all cross-sectional dimensions than Samples A, but yet Samples B exhibits two orders of magnitude more EM-induced plasticity.

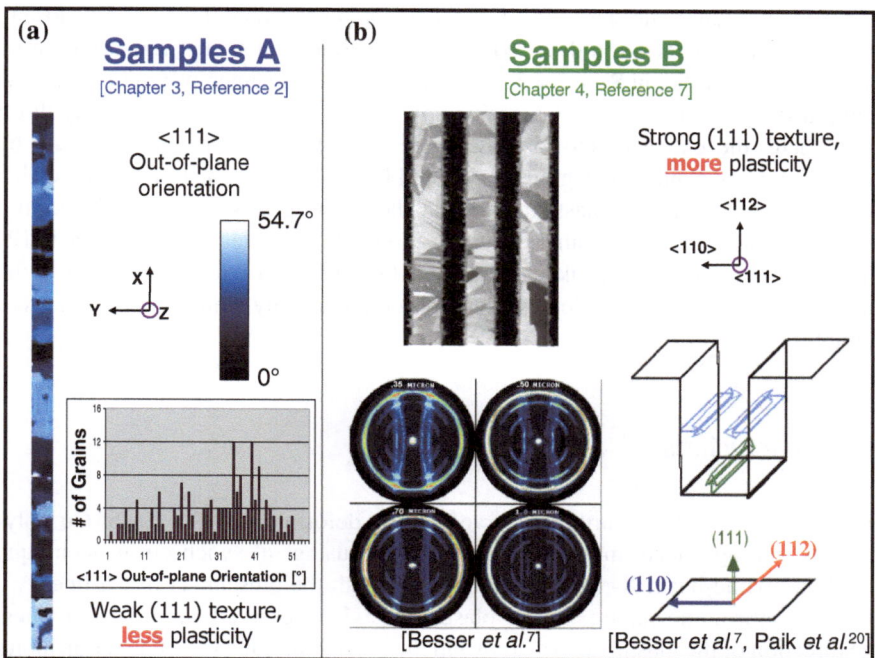

Fig. 4.9 Samples A versus B comparison: the texture correlation; **a** Samples A (Cu lines reported in Chap. 3, as well as in the Ref. [2]) have weak (111) texture, and "less plasticity," whereas; **b** Samples B (Cu lines reported in this chapter, similar to the Cu lines in Ref. [7]) have strong (111) texture, and "more plasticity"; Illustration of strong preferred in-plane orientation of (111) grains: $\langle 110 \rangle$ to the sidewalls, $\langle 112 \rangle$ along the direction of length of the lines (Courtesy of Besser et al. [7])

Figure 4.9 is a comparison summary of the known information about the Cu lines in Samples A versus Samples B. First, Samples A (Fig. 4.9a) shows a weak (111) texture, and we found the extent of EM-induced plasticity in the order of $\rho \sim 10^{12}$–10^{13}/m². Subsequently, Samples B were found with $\rho \sim 10^{15}$/m² after similar electromigration conditions—a significantly larger amount of EM-induced plasticity. Besser et al. [7] suggested that Samples B have the typical strong (111) texture.

This observation of significantly larger EM-induced plasticity in Samples B compared to that of Samples A, thus, is consistent with our earlier observation, especially detailed in Chap. 3, that the occurrence of plastic deformation in a given grain can be strongly correlated with the availability of a $\langle 112 \rangle$ direction of the crystal in the proximity of the direction of the electron flow in the line (within an angle of 10°). In $\langle 111 \rangle$ out-of-plane oriented grains in a damascene interconnect scheme, the crystal plane facing the sidewall tends to be a $\{110\}$ plane, so as to minimize the interfacial energy [7, 19, 20]. Therefore, it is deterministic rather than probabilistic that the $\langle 111 \rangle$ grains will have a $\langle 112 \rangle$ direction nearly parallel to the direction of electron flow or the direction of the length of the line. This is illustrated in Fig. 4.9b.

In Samples B, most grains are $\langle 111 \rangle$ in out-of-plane orientation (such as shown in the FIB mapping in Fig. 4.9b), and thus prefer energetically to have the $\langle 110 \rangle$

directions normal to the sidewalls, thus causing a $\langle 112 \rangle$ direction to be very close to the direction of the electron flow. When this condition is met, our proposed correlation, discussed earlier in Sect. 3.4.5, but more in detailed manner in Ref. [6], suggests that plasticity occurs in these Cu grains upon electromigration, and not only did it occur here, the extent of the plasticity here was rather extreme. Samples A, by comparison, have only a few grains that are $\langle 111 \rangle$ in out-of-plane orientation, which led to the occurrence of plasticity only in these few grains in the Cu lines after electromigration. In most other grains (i.e. non $\langle 111 \rangle$-oriented grains), a $\langle 112 \rangle$ direction of the Cu crystal is likely to be not in the direction of the electron flow of the lines. This is the reason plasticity was not observed in many grains in the Cu lines of Samples A.

4.5 Conclusions

In conclusion, we have further observed plastic deformation behavior of Cu polycrystals during electromigration experiments, using a synchrotron technique involving white-beam X-ray microdiffraction. With the present set of Cu lines, we further confirm the nature of the plastic behavior observed in these Cu lines: the direction of the plastic deformation and the availability of a $\langle 112 \rangle$ direction along the length of the lines as the rotation axis of the plastic deformation. Furthermore, we found that the extent of the electromigration-induced plasticity in this set of Cu lines was significantly larger compared to that of the previous study. We propose that the crystallographic texture of the Cu lines plays a primary role in controlling the plastic behavior of the interconnect lines. Strong (111) texture leads to high preference of $\langle 112 \rangle$ direction along the length of the line, and this subsequently leads to higher tendency for the grains to behave plastically in respond to electromigration stressing.

References

1. Vanasupa L, Joo YC, Besser PR et al (1999) Texture analysis of damascene-fabricated Cu lines by x-ray diffraction and electron backscatter diffraction and its impact on electromigration performance. J Appl Phys 85:2583–2590
2. Budiman AS, Tamura N, Valek BC et al (2006) Crystal plasticity in Cu damascene interconnect lines undergoing electromigration as revealed by synchrotron x-ray microdiffraction. Appl Phys Lett 88:233515
3. Valek BC, Bravman JC, Tamura N et al (2002) Electromigration-induced plastic deformation in passivated metal lines. Appl Phys Lett 81:4168–4170
4. Valek BC, Tamura N, Spolenak R et al (2003) Early stage of plastic deformation in thin films undergoing electromigration. J Appl Phys 94:3757–3761
5. Budiman AS, Tamura N, Valek BC et al (2004) Materials, technology and reliability for advanced interconnects and low-k dielectrics. Mat Res Soc Proc 812:345–350
6. Budiman AS, Tamura N, Valek BC et al (2006) Electromigration-induced plastic deformation in Cu Damascene interconnect lines as revealed by synchrotron x-ray microdiffraction. Mat Res Soc Proc 0914-F06-01–0914-F06-05

7. Besser P, Zschech E, Blum W et al (2001) Microstructural characterization of inlaid copper interconnect lines. J Elec Matls 30:320–330
8. Lingk C, Gross ME, Brown WL (2000) Texture development of blanket electroplated copper films. J Appl Phys 87:2232–2236
9. Harper JM, Colgan EG, Hu CK et al (1994) Materials issues in copper interconnections. Mat Res Soc Bull 23:23–29
10. Besser PR, Sanchez JE, Field DP (1997) Proceedings of the advanced metallization and interconnect systems for ULSI applications in 1996, vol 89. Materials Research Society, Warrendale, PA
11. Lingk C, Gross ME, Brown WL, Siegrist T, Coleman E, Lai WY-C, Miner JF, Ritzdorf T, Turner J, Gibbons J, Klawuhn E, Wu G, Zhang F (1999) Advanced metallization conference in 1998 (AMC 1998), vol 73. Materials Research Society, Warrendale, PA
12. Lingk C, Gross ME, Brown WL (1999) X-ray diffraction pole figure evidence for (111) sidewall texture of electroplated Cu in submicron damascene trenches. Appl Phys Lett 74:682–684
13. Besser PR, Sanchez JE, Field DP, Pramanick S, Sahota K (1997) Advanced metallization for ULSI applications. Mat Res Soc Proc 473:217
14. Besser PR, Joo YC, Winter D et al (1999) Mechanical stresses in aluminum and copper interconnect lines for 0.18 μm logic technologies. Mat Res Soc Proc 563:189
15. Besser PR (1999) Stress-induced phenomena in metallization. In: AIP conference proceedings, vol 491, p 229
16. Zschech E, Besser PR (2000) Microstructure characterization of metal interconnects and barrier layers: status and future. In: Proceedings of the international interconnect technology conference, vol 233. p 235
17. Venkatasen S, Gelatos A, Misra V, Smith B, Islam R, Cope J, Wilson B, Tuttle D, Cardwell R, Anderson S, Angyal M, Bajaj R, Capasso C, Crabtree P, Das S, Farkas J, Fillipiak S, Fiordalice B, Freeman M, Gilbert P, Herrick M, Jain A, Kawasaki H, King C, Klein J, Lii T, Reid K, Saaranen T, Simpson C, Sparks T, Tsui P, Venkatraman R, Watts D, Wietzman E, Woodruff R, Yang I, Bhat N, Hamilton G, Yu Y (1997) In: IEEE International Electron Device Meeting Digest, vol 769. IEEE, Piscataway, NY)
18. Licata T, Okazaki M, Ronay M et al (1995) Dual damascene Al wiring for 256M DRAM. In: Proceedings of the VLSI multilevel interconnection conference
19. Sanchez JE, Besser PR (1998) Proceedings of the international interconnect technology conference, vol 233. IEEE, Piscataway, NY
20. Paik JM, Park KC, Joo YC (2004) Relationship between grain structures and texture of damascene Cu lines. J Elec Matls 33:48–52
21. Diebold A, Goodall RK (1999) Interconnect metrology roadmap: status and future. In: Proceedings of the international interconnect technology conference, San Francisco, 24–26 May 1999
22. Rhee SH, Du Y, Ho PS (2003) Thermal stress characteristics of Cu/oxide and Cu/low-k submicron interconnect structures. J Appl Phys 92:3926–3833
23. Paik JM, Park H, Joo YC et al (2005) Effect of dielectric materials on stress-induced damage modes in damascene Cu lines. J Appl Phys 97:104513
24. Fayolle M, Passemard G, Assous M et al (2002) Integration of copper with an organic low-k dielectric in 0.12-μm node interconnect. Micoelectron Eng 60:119–124
25. Filippi RG et al (2004) Thermal cycle reliability of stacked via structures with copper metallization and an organic low-k dielectric. In: 42nd annual IEEE international reliability physics symposium proceedings, pp 61–67
26. Shen YL (2006) Effects of dielectric thermal expansion and elastic modulus on the stress and deformation fields in copper interconnects. Mat Res Soc Proc 0914-F04-01–0914-F04-10
27. Tamura N, MacDowell AA, Spolenak BC et al (2003) Scanning x-ray microdiffraction with submicrometer white beam for strain/stress and orientation mapping in thin films. J Sync Rad 10:137–143

Chapter 5
Industrial Implications
of Electromigration-Induced Plasticity
in Cu Interconnects: Plasticity-Amplified
Diffusivity

Abstract The theoretical analysis of the diffusion in interconnects during the electromigration is performed in this chapter. Two main diffusion paths are compared: grain boundaries and electromigration induced dislocations. The electromigration induced dislocations are formed due to the crystal bending, described in the previous chapters, and form a short diffusion path in the direction of the current. These two possible diffusion paths can lead to the significantly different device failure time dependence on the current density. This can lead to an important implication for the way device lifetime/reliability is assessed.

Keywords Electromigration · Interconnect · Diffusion path · Electron wind · Reliability

5.1 Introduction

The mass transport of Cu during electromigration (EM) testing is typically dominated by interface diffusion. If a mechanism other than interface diffusion begins to affect the overall transport process, then the effective diffusivity, D_{eff}, of the electromigration process would deviate from that of interface diffusion only. This would have fundamental implications. We have preliminary evidence that this might be the case. Plastic deformation has been observed in damascene Cu interconnect test structures during an in situ electromigration experiment (described in Chaps. 3 and 4, as well as reported in the literature [1–3]), using a synchrotron technique of white beam X-ray microdiffraction as discussed in Chapter. A particular direction of the plasticity has been observed consistently over a few different sets of Cu, as well as Al interconnect lines [4, 5]. This leads to a specific configuration of edge dislocations that allow additional global transport of atoms during electromigration. We propose that this effect manifests itself in the deviation of the observed current density exponent, n, from its expected value of unity, in Black's Law. This has important industrial practical implications as n lies at the heart of the

© The Author(s) 2015
A.S. Budiman, *Probing Crystal Plasticity at the Nanoscales*,
SpringerBriefs in Applied Sciences and Technology,
DOI 10.1007/978-981-287-335-4_5

reliability assessment of the interconnect device. In this chapter we report its implications for electromigration device lifetime prediction.

5.2 Background

As interconnects are aggressively scaled, current density continues to increase, thereby accelerating the electromigration degradation processes. One way the electromigration current could introduce extra damage is through electromigration-induced plasticity—a phenomenon that has been observed recently both in Al [4, 5] as well as—as has been described in Chap. 3 and—in Cu interconnects. This plasticity could have very important implications both in the efforts to improve electromigration reliability as well as in the development of methodologies for electromigration reliability assessment.

Plastic deformation was observed in metallic interconnect test structures [1–5] during in situ electromigration experiments and before the onset of visible micro-structural damage (voids, hillock formation). It has been shown using a synchrotron technique with white beam X-ray microdiffraction—as has been described in Chap. 2, as well as discussed in detail in literature [6]—that almost as soon as the electromigration current is turned on, grains start to deform plastically. The extent of this electromigration-induced plasticity in Cu test structures is dependent on the width of the interconnect lines (Chap. 3). In wide lines, plastic deformation manifests itself as grain bending and the formation of subgrain structures, while only grain rotation is observed in the narrower lines. More importantly, it has also been suggested that the grain texture of the Cu line play a primary role in the occurrence and, thus, the extent of the electromigration-induced plasticity—in a much bigger way than its dependence on linewidth (Chap. 4).

Furthermore, the grain bending and the subgrain structure formation were observed not in any random direction, but always in the direction across the width of the lines [1–5]. Upon further investigation, it was also found that a particular slip system in an FCC crystal is responsible for the kind of plasticity observed in the Cu grains, and that particular slip system always has the $\langle 112 \rangle$ line direction almost coinciding with the direction of the electron flow (within 10°). This correlation has been consistently observed over multiple sets of Cu interconnect lines fabricated by two different manufacturers (Chaps. 3 and 4), and has been discussed in detail in Sect. 3.4.5.

5.2.1 Dislocation Cores as Fast Diffusion Paths in Metallic Interconnects

The observed plasticity described above leads to a concentration of same-sign edge dislocations with cores running along the direction of electron flow such as illustrated in Fig. 5.1.

Fig. 5.1 Schematic of a grain
containing same-sign edge
dislocations with cores
running along the direction of
the electron flow in the
interconnect line

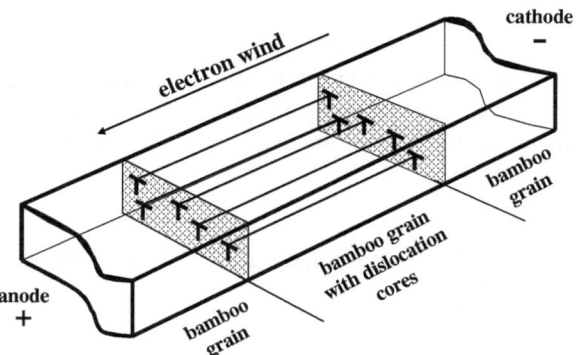

When this configuration of same-sign edge dislocations extends through grains
along the full length of the interconnect lines, they allow an additional path for
diffusion of atoms from one end of the interconnect line to the other. Dislocation
cores are, in general, already recognized as fast diffusion paths [7], but in this
configuration especially, their contribution to the overall migration of atoms from
the cathode to the anode end of the line is even more pronounced. Furthermore,
when the concentration of these dislocations reaches a high enough level, their
contribution to the overall effective diffusivity (D_{eff}) can no longer be neglected. We
write the effective diffusivity as

$$D_{eff} = \frac{\delta}{h} D_{int} + \rho a_{core} D_{core} ,\qquad(5.1)$$

where a_{core} and D_{core} are the cross-sectional area and diffusivity of a dislocation
core, respectively, ρ is the dislocation density, and δ, h and D_{int} are the effective
interface diffusion thickness, the height of the line and the diffusivity of the
interface, respectively. In this equation, the diffusivities are described in the usual
way by $D = D_o \exp(-E_A/kT)$ where E_A is the activation energy, D_o is a constant, and
k is the Boltzmann's constant.

Diffusion along dislocation cores ("pipe diffusion") has been commonly included in models of diffusion-controlled deformation in bulk materials [8]. Suo [9]
considered the motion and multiplication of dislocations under the influence of an
electric current in a conductor line, and suggested that electromigration-driven
dislocation multiplication could itself lead to dislocation densities high enough to
affect electromigration degradation processes. Oates [10], however, did not see any
diffusivity effects that could be attributed to dislocations in his experimental study.
Baker et al. [7] through their experimental study of nanoindented Al lines
(width = 1 μm, mean grain size = 1.1 μm) showed that the effect of a dislocation
density of $10^{16}/m^2$ is comparable to diffusion through a grain boundary. These
studies all essentially suggest that if the dislocation density is sufficiently high, it
may affect the overall electromigration degradation processes in metallic interconnects, and thus could have fundamental implications.

Chapters 3 and 4 have given a key piece of experimental evidence that opens up the very possibility that such a high dislocation density is present in the Cu test structures undergoing electromigration. This chapter thus concerns itself with its important implications for the elecromigration degradation processes, as well as, and more specifically, for the electromigration reliability assessment methodologies.

5.2.2 Electromigration Reliability Assessment Methodology: Black's Law

Electromigration is a major reliability concern in the advanced microelectronics industry. This is due to the aggressive scaling of interconnect dimensions andintroduction of new materials and processing schemes leading to even more challenges in guaranteeing interconnect robustness against electromigration failure. Understanding the fundamental relationship between electromigration in interconnect lines and parameters of materials and processing that make up the interconnect structures, is thus very important.

In addition, it is also useful, especially for industry, to accurately assess the lifetime of the device under electromigration conditions. This involves taking data under accelerated conditions (i.e. high temperatures and current densities) and scaling it back to the device operational conditions. This electromigration lifetime extrapolation is commonly based on the well-established Black's Law [11]. This methodology for extrapolations determines whether a technology is sufficiently reliable against electromigration failure for a given specification, or whether further optimization in process and/or design is needed.

Black's Law expresses the median time to failure (MTF), or the 50th percentile fail time of a failure population, as:

$$MTF = A\left(\frac{1}{j}\right)^n \exp\left(\frac{E_A}{kT}\right), \tag{5.2}$$

where A is an empirically-determined constant, j is the current density, n is the current density exponent, E_A is the activation energy to electromigration failure, k is the Boltzmann's constant and T is the absolute temperature.

The figure of merit for electromigration reliability is the use current density (current density at device operational conditions), which is commonly denoted as j_{use} or j_{max}, and represents the maximum current density the interconnect system can maintain while still guaranteeing a certain failure rate over a certain amount of operation time at use conditions. Therefore the current density exponent, n, is crucial, as the extrapolated failure time (i.e. device lifetime at use/operational conditions) is very sensitive to it. It has been established through studies on Al interconnects that void-growth-limited failures are represented by a current density exponent of one [12] ($n = 1$), while void-nucleation-limited failures are represented by a current density exponent of 2 ($n = 2$) [13, 14]. These concepts appear to be applicable to the case of Cu interconnects [15].

However, it has been widely observed that current density exponent, n, is usually found in real cases to be >1 (as opposed to $n = 1$ for the prevailing model of void growth limited failures) [12, 16]. This suggests that there is an extra dependency on j, under accelerated test conditions. This extra dependency has been attributed to the effect of Joule heating [16–18]. Joule heating is the process by which the passage of an electric current through a conductor releases heat. It is caused by interactions between electrons that make up the body of the conductor. At higher j (accelerated/ test conditions), more electrons are passing in the interconnect line, causing more heating, and thus a higher temperature (making it even higher than the accelerated/ test temperature), leading to amplification of electromigration diffusion in the interconnect lines, and thus earlier failure events. This manifests in the Black's Law as an extra dependency on j, or in other words, the deviation of n from unity (and/or from $n = 2$, for that matter).

5.3 Plasticity-Amplified Diffusion in Electromigration

We now come back to the configuration of high-density same-sign edge dislocations (Fig. 5.1) that has been observed, more specifically, in the Cu interconnect lines discussed in Chap. 4. In Fig. 5.2, we reproduce Fig. 4.5 from the last chapter to reiterate this observation, as well as the fact that these configurations are observed consistently across grains throughout the segment of the Cu interconnect lines studied. In this Sect. 5.3, we discuss the implications of these observations.

This configuration allows an alternative diffusion path—in addition to the dominant interface diffusion—in Cu interconnect lines in the event of bias by electromigration. When under certain circumstances, this diffusion gets activated, and its magnitude approaches that of the interface diffusion (thus "plasticity-amplified" diffusion), the line's electromigration behavior could deviate significantly from its interface-dominated behavior. This could also thus influence the way the reliability engineers assess the failure times and thus the lifetime of the device.

From Eq. 5.1, we notice there are at least two important requirements for this configuration of same-sign edge dislocations to influence electromigration significantly. First, only when there is high enough density (ρ) of these dislocations that the second term in Eq. 5.1 (i.e. the core diffusion contribution to the overall effective diffusion of electromigration) can become no longer negligible (We discuss this in the coming Sect. 5.3.1). Secondly, as the Cu line consists of mostly bamboo grains, the effective D_{core} in Eq. 5.1 still depends to some extent on the grain boundary diffusion. When grain boundaries inhibit the atomic transport from one dislocation core to another, the effective D_{core} could become very small and thus make the whole second term negligible. We cover this in the next Sect. 5.3.2. Only when these two conditions are satisfied, can we expect a truly global effect of plasticity-amplified electromigration diffusion in Cu interconnect lines.

Fig. 5.2 a Streaking and/or splitting of Cu Laue diffractions spots throughout a segment of the line; **b** Dislocations were found with cores lining up with the direction of the electron flow in the line (consistent with earlier observation discussed in Sect. 3.4) across grains throughout the length of the segment of the line observed (*Grain map is estimated based on streaking observation)

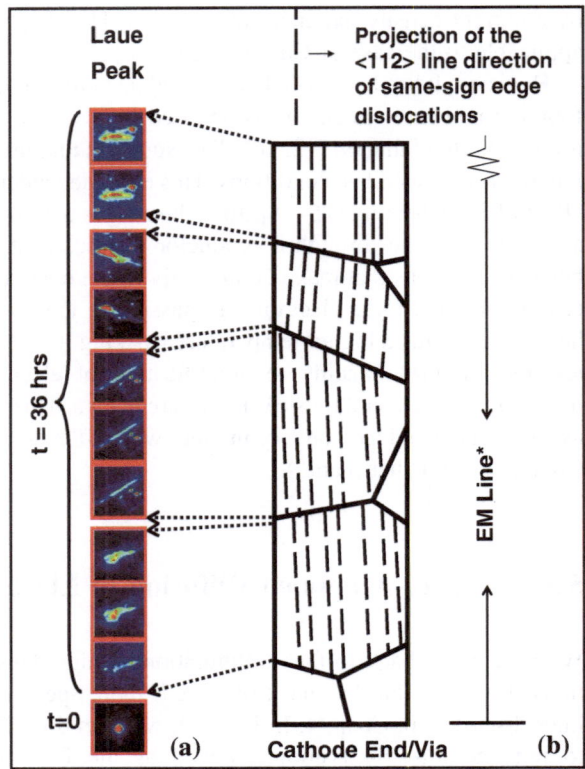

5.3.1 Density of Core Dislocations (ρ_{core}): Extent of Plasticity

We have established, in Chap. 4, that dislocations with cores running along the electron flow direction and densities in the order of $10^{15}/m^2$ are present in the Cu lines undergoing electromigration (accelerated test conditions) for 36 h. Figure 5.3 is a calculated (not experimentally observed) comparison of diffusivities as a function of temperature between the interface diffusion path and those of dislocation cores of various densities in Cu interconnect lines (from $10^{12}/m^2$, to $10^{15}/m^2$, to $10^{17}/m^2$) when each diffusion mechanism is assumed to act alone. The diffusivities here are described in the usual way by $D = D_o \exp(-E_A/kT)$ where E_A is the activation energy, D_o is a constant, and k is the Boltzmann's constant, and are calculated based on diffusion coefficient values in the literature [8, 19] for Cu interconnect lines, and for the interconnect dimensions as used in the study described in Chap. 4 (summarized in Table 5.1).

The level of ρ observed in Chap. 4 ($\rho_{GND} = \rho_{core} = 10^{15}/m^2$) is illustrated as the solid line in Fig. 5.3, and it shows here that the level of diffusion is within the same order of magnitude with that of interface (the dotted line) at the test conditions ($T = 300$ °C or $1000/T = 1.75/K$). Thus, $\rho_{core} = 10^{15}/m^2$ is just above the threshold

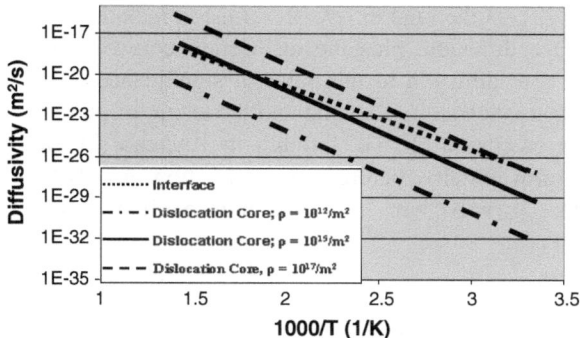

Fig. 5.3 Comparison of diffusivities as a function of temperature between the interface diffusion path and those of dislocation cores of various densities in Cu interconnect lines (from $10^{12}/m^2$, to $10^{15}/m^2$, to $10^{17}/m^2$) when each diffusion mechanism is assumed to act alone; Diffusivities were calculated using values summarized below (Table 5.1)

Table 5.1 Values used to determine diffusions in Cu interconnects as a function of temperature (Fig. 5.3). D_o is the pre-exponential constant and E_A is the activation energy. The subscripts *int* and *core* refer to interface and core diffusions, respectively. The δ is the effective interface diffusion thickness, h is the thickness of the Cu lines, and a_{core} is the area of a dislocation core

Variable	Value	Reference/Remarks
$\delta D_{o,int}$	3.4×10^{-19} m^3/s	Based on SiN/Cu, Ref. [19]
h	0.2 μm	Section 4.3
$E_{A,int}$	0.91 eV	Based on SiN/Cu, Ref. [19]
$a_{core}D_{o,core}$	1.0×10^{-24} m^4/s	For copper, Ref. [8]
$E_{A,core}$	1.21 eV	For copper, Ref. [8]

of dislocation density necessary for the dislocation core diffusion to be on par with the interface diffusion. In other words, at this dislocation density we can expect the contribution of dislocation cores to the overall/effective diffusivity in the Cu line during accelerated electromigration to be at least the same order of magnitude as interface diffusion.

It is to be noted, however, that at temperatures below 100 °C, it takes ρ_{core} in the order of $10^{17}/m^2$ (the dashed line in Fig. 5.3) for the effect of dislocation cores to be as significant. These lower temperatures correlate with the typical use/operational conditions of the interconnects. The typical initial (as fabricated) dislocation density in Cu/metallic lines was taken to be $10^{12}/m^2$ (following Ref. [7]), and the corresponding diffusivity is as shown by the dashed-dotted line in Fig. 5.3.

It is therefore reasonable to propose that the contribution from the dislocation core diffusion (the second term in Eq. 5.1) could no longer be neglected in the Cu lines studied in Chap. 4 during electromigration at accelerated test conditions. This contribution would enhance the electromigration diffusion, or in other words, the total EM flux (J_{EM}), as the total or overall diffusion includes the existing, usually-dominant interface diffusion, plus the dislocation core or "pipe" diffusion. The increase in this core diffusion to the point of significance in the overall electro-migration diffusion is related to the kind and the extent of plasticity induced by the electromigration process itself, i.e. through the increase in the core dislocation density, ρ_{core}—from initially, before electromigration, $\rho = 10^{12}/m^2$ to after some electromigration, $\rho_{core} = 10^{15}/m^2$—in Cu interconnect lines, as observed in Chap. 4.

5.3.2 Effect of Grain Boundary Diffusion: Effective D_{core}

Concentration of dislocation cores running along the direction of the length of the line in the grains in interconnect lines (such as illustrated in Fig. 5.1) during electromigration has been observed. It has been proposed, in the last subsection, that at a dislocation density, $\rho_{core} > 10^{15}/m^2$, such configuration of dislocations may lead to an additional path for diffusion in Cu lines during electromigration (in addition to the dominant upper/interface diffusion). However, for such a configu-ration to cause a real, global effect in the kinetics along the Cu lines, there is another requirement. Continuous diffusion paths (across grains) must be available for atoms to transport from the cathode end to the anode end of the lines. Considering the mostly bamboo grain structure that our interconnect lines have (Sect. 3.4.1), this requires consideration of grain boundary diffusivity, as atoms eventually hit grain boundaries and have to travel some distance in the grain boundary region before finding another set of dislocation cores (belonging to the neighboring grain) to continue their travel to the other end of the line. This is illustrated in Fig. 5.4.

In Eq. 5.1, the parameter D_{core} takes into consideration only the diffusion along the dislocation cores for the overall length of specimen of interest, which would be

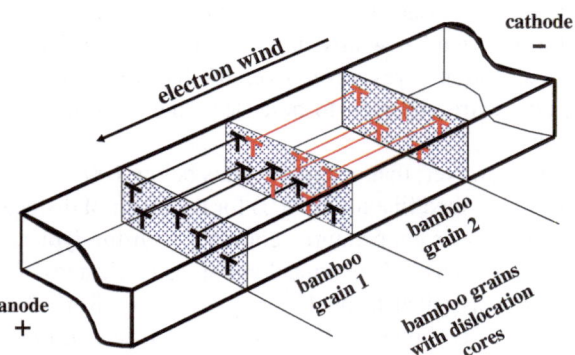

Fig. 5.4 Illustration of bamboo grains with dislocation cores running along the direction of the electron flow in the line under electromigration bias; Cores from one grain must end at the grain boundaries, and thus atoms traveling through them, must diffuse in grain boundary regions, before finding another set of dislocation cores in the next grains

true only for a single crystal Cu along the full length of the line. Clearly, our case, such as illustrated in Fig. 5.4, is not a single crystal and thus the effect of grain boundary diffusion must be considered. In the present Sect. 5.3.2, we study quantitatively the impact of grain boundary diffusivity on the overall dislocation core diffusion $\left(D_{core}^{effective}\right)$.

We first consider the grain boundary region as illustrated in Fig. 5.5, and suggest a relation between the mean distance, R, that atoms need to travel in the grain boundary region before finding another set of dislocation cores belonging to the next grain, and the dislocation density, ρ_{core}, in the Cu lines induced by the electromigration process itself (Eq. 5.3). As can be expected, R is inversely related to ρ_{core} (or in other words, the more dislocation cores in the cross-section of the Cu lines, the smaller the diffusion distance in the grain boundary region).

It is clear from Fig. 5.5 that:

$$R = \frac{l}{\sqrt{2}}; \quad \rho_{core} = \frac{1}{l^2},$$

and thus,

$$R = \frac{1}{\sqrt{2\rho_{core}}} \tag{5.3}$$

Next, we consider the effective diffusion along a hypothetically continuous dislocation core, as well as the actual diffusion along the dislocation core and along the grain boundaries connecting dislocation cores in one grain with those in

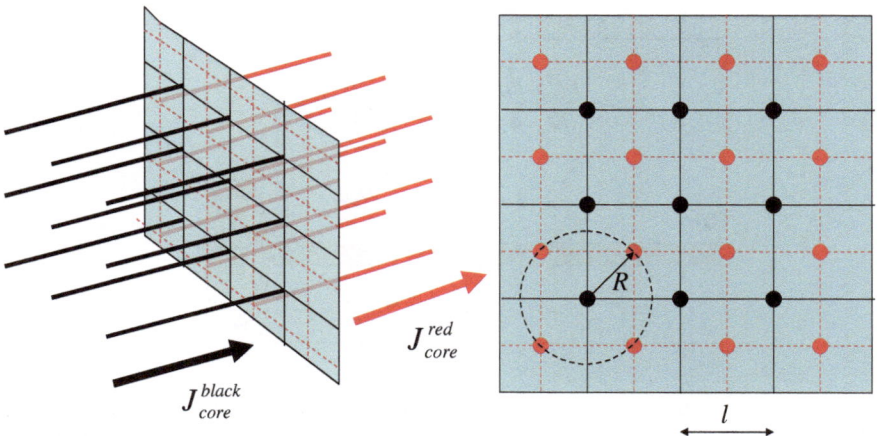

Fig. 5.5 Illustration of the grain boundary region of two bamboo grains with each dislocation cores running along the direction of the electron flow in the line under electromigration bias; Atoms traveling along the cores of the first grain (*black-colored*) must diffuse in grain boundary regions for the distance, R, before finding another set of dislocation cores in the next grains (*red-colored*)

another, for a grain size, L. For the effective core diffusion along the hypothetical dislocation core of length L, the flux, J_{eff}, can be expressed as

$$J_{eff} = -\frac{D_{core}^{eff} c}{kT} \frac{d\mu}{dx},$$ (5.4)

where $c = 1/\Omega$ (c = concentration of diffusing species; Ω = atomic volume), μ = chemical potential, x is the axis of the diffusion direction along the dislocation core, and D_{core}^{eff}, k and T have been defined before. This is illustrated in Fig. 5.6a.

The flux along a dislocation core, q_{core}, is simply

$$q_{core} = J_{eff} a_{core},$$ (5.5)

as is clear from Fig. 5.6a, where a_{core} is the cross-sectional area of a dislocation core ($=\pi r_{core}^2$). The fluxes in Fig. 5.6b consist of:

1. flux along the dislocation core, q_{core}
2. flux along the grain boundary region, q_{gb}

We now consider the diffusion along a dislocation core of length, L, as driven by the chemical potential difference, $\Delta\mu_1$, as follow

$$J_1 = -\frac{D_{core}}{kT\Omega} \frac{\Delta\mu_1}{L},$$ (5.6)

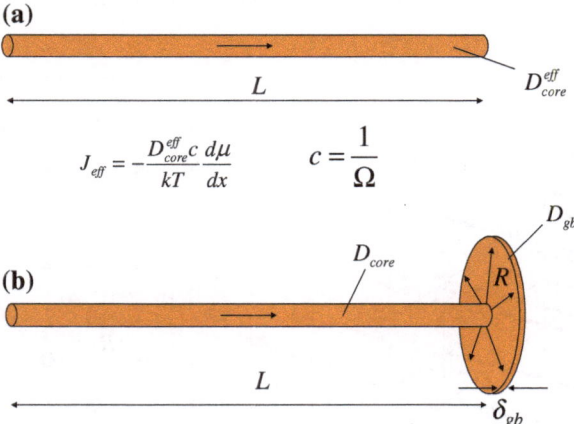

(a)

$$J_{eff} = -\frac{D_{core}^{eff} c}{kT} \frac{d\mu}{dx} \qquad c = \frac{1}{\Omega}$$

(b)

Fig. 5.6 Illustration of **a** a hypothetically continuous dislocation core diffusion along a specified diffusion distance, L, and the basic formula expressing the effective flux, J_{eff}, and its effective diffusivity, D_{core}^{eff}, and **b** the model of the combined effects of dislocation cores and grain boundaries along a normalized diffusion distance; δ_{gb} is the effective thickness of the grain boundary; These two illustrations ((**a**) and (**b**)) define D_{core}^{eff}

which leads to

$$q_{core} = J_1 a_{core} = \left(-\frac{D_{core}}{kT\Omega}\frac{\Delta\mu_1}{L}\right)(\pi r_{core}^2). \tag{5.7}$$

Along the grain boundary region, which can be modeled as a donut-shaped disc (Fig. 5.7) with disc thickness, δ_{gb}, and inner diameter, r_{core}, and outer diameter, R, the diffusion can be described as follow

$$q_{gb} = J_2(2\pi r \delta_{gb}), \tag{5.8}$$

or equivalently,

$$q_{gb} = 2\pi r \delta_{gb}\left\{-\left(\frac{D_{gb}}{kT\Omega}\right)\frac{d\mu}{dr}\right\}, \tag{5.9}$$

where q_{gb} is the flux along the grain boundary, and J_2 is the grain boundary flux per area, while the variables μ and r are the chemical potential and radius/distance of diffusion from the center of the disc, respectively. However q_{gb} is not a function of $r\,(\neq f(r))$, and thus should be a constant. Consequently, by rearranging Eq. 5.9, we find

$$-q_{gb}\frac{kT\Omega}{2\pi\delta_{gb}D_{gb}} = r\frac{d\mu}{dr} = const. \tag{5.10}$$

Fig. 5.7 Illustration of the diffusion along the grain boundary region with indications of the parameters (R, r_{core}, q_{gb} and δ_{gb}) later used in the analyses

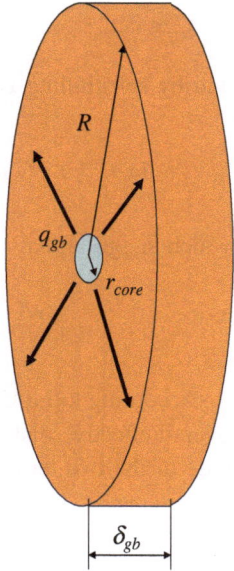

Integrating and further rearranging leads to

$$\Delta\mu_2 = -q_{gb}\frac{kT\Omega}{2\pi\delta_{gb}D_{gb}}\ln\left(\frac{R}{r_{core}}\right),\tag{5.11}$$

where $\Delta\mu_2$ is simply the integrated chemical potential difference for diffusion along the grain boundary region from $r = r_{core}$ to $r = R$.

In the combined diffusion along the actual dislocation core and along the grain boundary region (Fig. 5.6b), mass conservation requires

$$q_{core} = q_{gb},\tag{5.12}$$

which after substituting Eqs. 5.7 and 5.11 would lead to

$$-\pi r_{core}^2\frac{D_{core}}{kT\Omega}\frac{\Delta\mu_1}{L} = -\Delta\mu_2\frac{2\mu\delta_{gb}D_{gb}}{kT\Omega\ln(R/r_{core})}.\tag{5.13}$$

Obviously, the combined chemical potential, $\Delta\mu$, can be described as follow

$$\Delta\mu = \Delta\mu_1 + \Delta\mu_2.\tag{5.14}$$

Combining Eqs. 5.13 and 5.14, we find

$$\Delta\mu_1 = \left\{\frac{2\delta_{gb}D_{gb}L}{2\delta_{gb}D_{gb}L + r_{core}^2 D_{core}\ln(R/r_{core})}\right\}\Delta\mu.\tag{5.15}$$

To derive the effective diffusivity of the core diffusion, D_{core}^{eff}, we rewrite Eq. 5.7

$$q_{core} = -\pi r_{core}^2\frac{D_{core}}{kT\Omega}\frac{\Delta\mu_1}{L},$$

and by substituting Eq. 5.15, we find

$$q_{core} = -\pi r_{core}^2\frac{D_{core}}{kT\Omega}\left\{\frac{2\delta_{gb}D_{gb}L}{2\delta_{gb}D_{gb}L + r_{core}^2 D_{core}\ln(R/r_{core})}\right\}\frac{\Delta\mu}{L},\tag{5.16}$$

which suggests that

$$D_{core}^{eff} = D_{core}\left\{\frac{2\delta_{gb}D_{gb}L}{2\delta_{gb}D_{gb}L + r_{core}^2 D_{core}\ln(R/r_{core})}\right\}.\tag{5.17}$$

According to Eq. 5.17 therefore, the influence of grain boundary diffusion on the overall/effective dislocation core diffusivity, D_{core}^{eff}, depends on the relative magnitude of the two terms in the denominator in the Eq. 5.17. If

$$2\delta_{gb}D_{gb}L \gg r_{core}^2 D_{core}\ln(R/r_{core}),\tag{5.18}$$

then as evident from Eq. 5.17, D^{eff}_{core} degenerates into simply D_{core}, or in other words, there is very little influence of the grain boundary diffusion in the overall scheme in Fig. 5.6b. If the reverse is true, D^{eff}_{core} will be much smaller than D_{core}, in which case it is clear that the grain boundary slows down significantly the overall diffusion in Fig. 5.6b.

It is evident from Table 5.3 that as it is, the first term is larger by at least 4 orders of magnitudes than the second. This would lead to the degeneration of D^{eff}_{core} into simply D_{core} in Eq. 5.17, which would suggest that a practically continuous pipe (dislocation core) diffusion path across the grains between the cathode end and the anode end of the lines is indeed available for atoms to transport in the Cu test structures under accelerated EM testing that we discuss in Chap. 4. This is shown in Fig. 5.8 (The red-colored/crossed D^{eff}_{core} line and the yellow-colored/buttoned D_{core} line are practically on top of each other).

An extreme would be to take $E_{A,gb}$ to be the E_A of lattice diffusion, which is 2.04 eV.[8] This is a much higher activation energy than that of the grain boundary. In this case, we show that the combined diffusivity would be dominated by such a slow diffusion in the hypothetical "grain boundary." The effective transport through dislocation cores in this case would be slowed down by close to 4 orders of magnitude due to the effect of the hypothetical grain boundary, such as shown in Fig. 5.8 (with the blue-colored plain line).

Table 5.2 Values used to determine influence of grain boundary diffusion on the overall transport of Fig. 5.6b. The diffusivities (D_{gb}, D_{core}) are described in the usual way by $D = D_o \exp(-E_A/kT)$ where E_A is the activation energy, D_o is the pre-exponential constant, and k is the Boltzmann's constant. The subscripts gb refers to grain boundary diffusion

Variable	Value	Reference/Remarks
T	300 °C = 573 K	Following T_{test} in Chap. 4
$E_{A,gb}$	1.08 eV	Refs. [8, 19, 20, 21]
$\delta_{gb}D_{gb}$	1.6×10^{-24} m^3/s	Calculated, Ref. [8]
L	1 μm	Estimated based on Chap. 4
r_{core}	0.25 Å	Ref. [8]
$E_{A,core}$	1.21 eV	Ref. [8]
$r^2_{core}D_{core}$	7.3×10^{-36} m^4/s	Calculated, Ref. [8]
ρ_{GND}	10^{15}/m^2	Observed in Chap. 4
R	22 nm	$R = 1/2\rho_{core}$

Table 5.3 Values of the two parameters/terms in Eq. 5.18 (or denominator of Eq. 5.17) calculated based on values listed in Table 5.2

Parameter/Term in Eq. 5.18	Value
$2\delta_{gb}D_{gb}L$	3.2×10^{-30} m^4/s
$r^2_{core}D_{core} \ln(R/r_{core})$	5.0×10^{-35} m^4/s

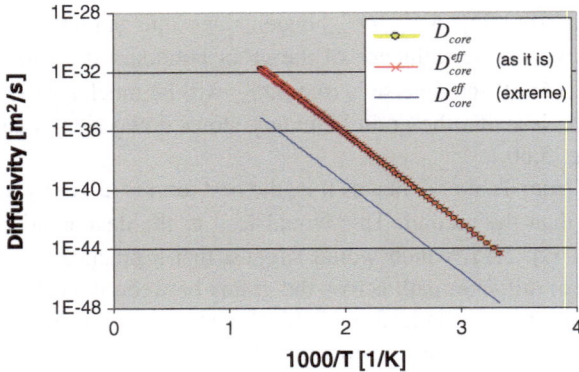

Fig. 5.8 Comparison of diffusivities as a function of temperature between D_{core} (only dislocation core diffusion, no grain boundary), D_{core}^{eff} (considering the effect of grain boundary; as it is—as shown in Table 5.3), and an extreme D_{core}^{eff} (considering the effect of grain boundary diffusion as if it is lattice diffusion). Diffusivities were calculated using values summarized in Table 5.2

It is therefore reasonable to propose that a fully continuous network of dislocation cores running along the direction of the length of the line, slowed only by less than 0.01 % (within the range of reported values of $E_{A,gb}$) by grain boundary diffusion, exists in the Cu interconnect lines studied during electromigration under accelerated test conditions in this study. This makes it a viable alternative for global transport of atoms in Cu interconnects under electromigration bias. Together, with $\rho_{GND} \sim 10^{15}/m^2$ observed in this study, and D_{core}^{eff} not much reduced by grain boundary diffusion, the second term in Eq. 5.1 (i.e. the contribution of the dislocation core diffusion) can indeed no longer be neglected. This means it will have important implications to the fundamental understanding of the electromigration degradation processes, as well as to the electromigration reliability assessment methodologies.

5.3.3 The Extra Dependency on J—The Plasticity Effect

If ρ should increase with j, then we will find that D_{eff} (of the electromigration process) should also increase with j. Consequently, there will be an extra electromigration flux, and thus an extra reduction in the time to failure of the device with increasing j. This is an extra dependency on j, which would manifest itself in the value of the current density exponent, n (in Black's equation), being >1.

The fact that n is usually found in real cases to be >1 (as opposed to $n = 1$ for the prevailing model of void growth limited failure) suggests that this extra dependency on j, especially at high temperatures of the test conditions, could be due to dislocation core diffusion. In other words, the higher n could be traced back to the higher level of plasticity in the crystal, and the closer n is to unity, the less plasticity must have influenced the electromigration degradation process.

Kirchheim and Kaeber [12] experimentally observed the *MTF* dependency on current density, j, in an Al conductor line, for a wide range of j, such as shown in Fig. 5.9 (the black-colored dots with error bars were the original data points). It clearly shows that at low current densities, the *MTF* data is best fit by $n = 1$ (straight dotted line), while at higher current densities, the *MTF* data is better fit by $n > 1$ (curved dotted line). Kirchheim and Kaeber [12] however suggested in their paper that these deviations occurring at higher current densities might have been caused by Joule heating.

Plasticity could just as likely be the source of such deviations of *MTF* dependency on j at high current densities. As j increases, plasticity also increases leading to increasingly higher electromigration fluxes (the dashed, arrowed lines in Fig. 5.9), and thus increasingly lower *MTF*, and therefore eventually a current density exponent, $n > 1$ has to be used to fit the failure time distribution.

However, under use conditions where the temperature is much lower (e.g. 100 °C), the level of ρ associated with such elevated diffusivity is almost impossible to reach, so that this plasticity-amplified diffusivity is associated only with the high temperature and high current density of the accelerated electromigration test. In other words, there is not likely to be much plasticity under use conditions, and thus the diffusivity is dominated only by interface diffusion, and consequently the *MTF* dependency on j should follow the $n = 1$ line. This is consistent with the observations of Kirchheim and Kaeber [12] (Fig. 5.9 shows data following $n = 1$ line at low reduced current densities, 0.2–1 MA/cm^2). This interpretation of the Kirchheim and Kaeber data is consistent with the physical model (void growth limited failure) which has also been observed through in situ electromigration studies on similar material by Zschech et al. [22].

It can be further stated that plasticity-amplified diffusivity is simply an extra mode of deformation under test conditions (which is not typically present under use conditions), and that its effect is wholly captured in the n value being greater than

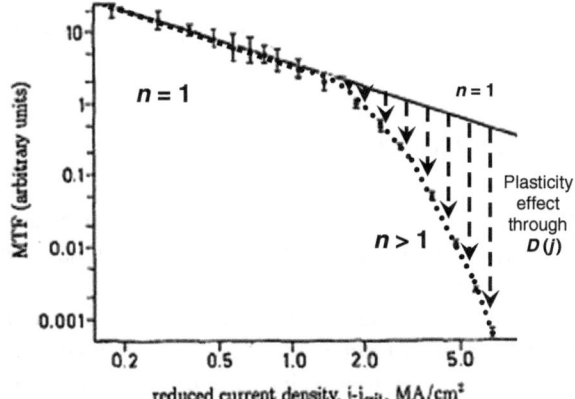

Fig. 5.9 Kirchheim and Kaeber's experimental *MTF* data as a function of reduced current density, $j - j_{crit}$ (all the solid features) —Courtesy of Ref. [12]; The *dotted* and *dashed lines* are added to lead to our argument

Fig. 5.10 Illustration of the danger of overestimation of device lifetime by using $n > 1$ (*red solid line*). Extrapolation using $n = 1$ (*blue dashed line*) is safer, and more likely to be closer to the actual device lifetime in use conditions. Kirchheim and Kaeber's experimental *MTF* data is again used for illustrative purposes (Courtesy of Ref. [12])

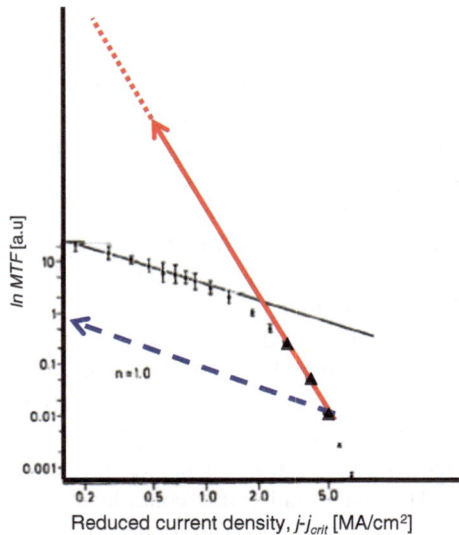

unity. This plasticity-inflated n could thus lead to inaccurate extrapolations of lifetimes under use conditions. This is illustrated in Fig. 5.10.

Figure 5.10 shows that if suppose we take the three *MTF* data points under accelerated j (the three solid black triangles), and based on these data points, we calculate n (which will be larger than 1), and then we use this n to extrapolate from the accelerated condition (high j) to the use condition (low j). That extrapolation is shown by the red solid line, and it clearly is an overestimation of the device's actual lifetime (approximated by the *MTF* data at low reduced current density).

To improve the accuracy of the reliability assessment of devices under use conditions, we therefore propose that the effect of plasticity has to be removed first from the electromigration lifetime equation. This can be done simply by insisting on $n = 1$ in our lifetime assessment (i.e., j_{max} calculation) which in most typical electromigration test conditions will result in a more conservative prediction of device lifetime, such as illustrated by the blue dashed line in Fig. 5.10.

5.4 Conclusions

We have used the synchrotron-based white beam Laue X-ray microdiffraction technique to investigate electromigration-induced plasticity in Cu interconnect structures undergoing electromigration testing. We discovered that the extent and configuration of dislocations in the Cu grains induced during this accelerated electromigration testing could lead to another competing electromigration diffusion mechanism in addition to interface diffusion. We have suggested that this plasticity effect can be correlated to the measured value of current density exponent, n, in Black's equation. We have observed that this correlation could then lead to an important implication for the way device lifetime/reliability is assessed.

References

1. Budiman AS, Hau-Riege CS, Besser PR et al (2007) Plasticity-amplified diffusivity: dislocation cores as fast diffusion paths in Cu interconnects. In: 45th annual IEEE international reliability physics symposium proceedings, Phoenix, 15–19 Apr 2007
2. Budiman AS, Tamura N, Valek BC et al (2006) Crystal plasticity in Cu damascene interconnect lines undergoing electromigration as revealed by synchrotron X-ray microdiffraction. Appl Phys Lett 88:233515
3. Budiman AS, Tamura N, Valek BC et al. (2006) Electromigration-induced plastic deformation in Cu damascene interconnect lines as revealed by synchrotron X-ray microdiffraction. Mat Res Soc Proc 914
4. Valek BC, Bravman JC, Tamura N et al (2002) Electromigration-induced plastic deformation in passivated metal lines. Appl Phys Lett 81:4168–4170
5. Valek BC, Tamura N, Spolenak R et al (2003) Early stage of plastic deformation in thin films undergoing electromigration. J Appl Phys 94:3757–3761
6. Tamura N, MacDowell AA, Spolenak BC et al (2003) Scanning X-ray microdiffraction with submicrometer white beam for strain/stress and orientation mapping in thin films. J Synchrotron Radiat 10:137–143
7. Baker SP, Joo YC, Knaub MP et al (2000) Electromigration damage in mechanically deformed Al conductor lines: dislocations as fast diffusion paths. Acta Mater 48:2199–2208
8. Frost HJ, Ashby MF (1982) Deformation-mechanism maps: the plasticity and creep of metals and ceramics. Pergamon Press, Oxford
9. Suo Z (1994) Electromigration-induced dislocation climb and multiplication in conducting lines. Acta Metall Mater 42:3581–3588
10. Oates AS (1996) Electromigration transport mechanisms in al thin-film conductors. J Appl Phys 79:163–169
11. Black JR (1967) Mass transport of aluminum by momentum exchange with conducting electrons. In: 6th annual IEEE international reliability physics symposium proceeding, Los Angeles, 6–8 Nov 1967
12. Kirchheim R, Kaeber U (1991) Atomistic and computer modeling of metallization failure of integrated circuits by electromigration. J Appl Phys 70:172–181
13. Korhonen MA, Borgesen P, Tu KN et al (1993) Stress evolution due to electromigration in confined metal lines. J Appl Phys 73:3790–3799
14. Lloyd JR (1991) Electromigration failure. J Appl Phys 69:7601
15. Hau-Riege CS, Marathe AP, Pham V (2002) The effect of line length on the electromigration reliability of Cu interconnects. In: Proceedings of the advanced metallization conference, vol 169
16. Schafft HA, Grant TC, Saxena AN et al (1985) Electromigration and the current density dependence. In: Reliability physics symposium, Orlando
17. Sigsbee RA (1973) Electromigration and metalization lifetimes. J Appl Phys 44:2533–2540
18. Bobbio A, Saracco O (1975) A modified reliability expression for the electromigration time-to-failure. Micoelectron Reliab 14:431–433
19. Gan D, Ho PS, Pang Y et al (2006) Effect of passivation on stress relaxation in electroplated copper films. J Mater Res 21:1512–1518

20. Cai B, Kong QP, Lu L et al (1999) Interface controlled diffusional creep of nanocrystalline pure copper. Scripta Mater 41:755–759
21. Dickenscheid W, Birringer R, Gleiter H et al (1991) Investigation of self-diffusion in nanocrystalline copper by NMR. Solid State Commun 79:683–686
22. Zschech E, Meyer MA, Langer E (2004) Effect of mass transport along interfaces and grain boundaries on copper interconnect degradation. In: MRS Proceedings, San Francisco, 12–16 Apr 2004

Chapter 6
Indentation Size Effects in Single Crystal Cu as Revealed by Synchrotron X-ray Microdiffraction

Abstract The observation of Laue peak streaking near small indentations in the (111) surface of a copper single crystal is described. The geometrically necessary dislocation (GND) density is computed from the μSXRD data for a different indentation depths. It is shown that GND density increases with decreasing indentation depth, which is in agreement with a revised Nix-Gao model. This finding supports that the indentation size effect is associated with geometrically necessary dislocations and related strain gradients.

Keywords μXRD · Indentation · Size effect · Smaller is stronger · Strain gradient

6.1 Introduction

The indentation size effect (ISE) has been observed in numerous nanoindentation studies on crystalline materials; it is found that the hardness increases dramatically with decreasing indentation size—a "smaller is stronger" phenomenon. Some have attributed the ISE to the existence of strain gradients and the geometrically necessary dislocations (GNDs). Since the GND density is directly related to the local lattice curvature, the Scanning X-ray Microdiffraction (μSXRD) technique has been utilized, which can quantitatively measure relative lattice rotations through the streaking of Laue diffractions. The synchrotron μSXRD technique we use—which was developed at the Advanced Light Source (ALS), Berkeley Lab—allows for probing the local plastic behavior of crystals with sub-micrometer resolution. Using this technique, we studied the local plasticity for indentations of different depths in a Cu single crystal. Broadening of Laue diffractions (streaking) was observed, showing local crystal lattice rotation due to the indentation-induced plastic deformation. A quantitative analysis of the streaking allows us to estimate the average GND density in the indentation plastic zones. The size dependence of the hardness, as found by nanoindentation, will be described, and its correlation to the observed lattice rotations will be discussed.

© The Author(s) 2015
A.S. Budiman, *Probing Crystal Plasticity at the Nanoscales*,
SpringerBriefs in Applied Sciences and Technology,
DOI 10.1007/978-981-287-335-4_6

6.2 Background

Modern devices are currently being aggressively scaled. Increasingly, the dimensions of these devices are at the sub-micrometer and nanometer scale. Although most of these devices are primarily functional and not mechanical, their reliability and lifetimes are often controlled by the mechanical properties of the materials that comprise the device. Thus, the creation of such small components requires a thorough understanding of the mechanical properties of materials at these small length scales. Furthermore, as specimens are reduced in size to the scale of the microstructure, their mechanical properties deviate from those of bulk materials. For example, in thin films—where only one dimension, the thickness, reaches the micron scale and below—the flow stress is found to be higher than its bulk value and becomes even higher as the film gets thinner. This thin film size effect is usually attributed to the confinement of dislocations by the substrate [1–3].

In nanoindentation experiments, where the length-scale of the deformation reaches the microstructural length-scale of the material, the governing relations between stress and strain deviate from the classical laws that apply to bulk materials. For crystalline materials, the hardness of a small indentation is usually higher than that of a large indentation. This indentation size effect (ISE) has been explained using the concept of geometrically necessary dislocations (GNDs) and strain gradients [4–18]. According to this picture, for a self-similar indenter, for example, a Berkovich-shape pyramidal indenter, the total length of GNDs forced into the solid by the indenter scales with the square of the indentation depth, while the volume in which these dislocations are found scales with the cube of the indentation depth; thus, the GND density (ρ_G) depends inversely on the indentation depth. The higher dislocation densities expected at smaller indentation depths lead naturally to higher strengths through the Taylor relation [19], and this leads to the ISE.

Characterizing the deformation zone below indentations has been a focus of many researchers [20–23]. In recent years, the use of focused ion beam (FIB) has enabled more accurate scanning electron microscope (SEM) imaging [24–26], as well as crystal orientation mapping using electron backscatter diffraction (EBSD) [27, 28] and transmission electron microscopy (TEM) [29, 30]. Scanning X-ray microdiffraction (μSXRD) using a focused polychromatic/white synchrotron X-ray beam can be used to determine the lattice rotation which is directly related to the local lattice curvature [31], strain gradients, and the GND density. Compared to many other techniques, such as EBSD and TEM, two advantages of μSXRD are non-destructive and a much larger detection depth. μSXRD has been described in a complete manner in the literature [32], and its capability as a local plasticity probe at small scales stems from the high brilliance of the synchrotron source, as well as the recent advances in X-ray focusing optics. This capability is also related to the continuous range of wavelengths in a white X-ray beam, allowing Bragg's law to be satisfied even when the lattice is locally rotated or bent, resulting in the observation of streaked Laue spots. μSXRD has been used in the study of the early stages of electromigration failure in metallic interconnect lines [33, 34], wherein lattice bending and GNDs are created by electromigration processes [33, 34].

The use of spatially resolved X-ray diffraction to measure local lattice rotations induced by indentation was pioneered by Ice's group [35–40]. In particular, they have provided a methodology for a clean measurement of lattice rotation associated with a 2 μm-deep Berkovich indentation [35, 36]. They demonstrated that [35, 36], at the center of one particular indentation side-face, the X-ray beam encounters a single rotation axis; at other positions, the X-ray beam may encounter multiple rotation axes, which complicates the resulting diffracted beams.

The present study builds upon and is complementary to this body of knowledge, and our primary focus is to compare ρ_G estimated through the observed lattice rotation to that expected from nanoindentation hardness results. Using μSXRD, we quantitatively study the streaking/broadening of Cu Laue peaks corresponding to different indentation depths, allowing us to estimate ρ_G in the individual indentation-induced plastic zones. Then, a revised Nix and Gao model [16, 17] is used to correlate the experimental hardness measurement with ρ_G. Finally, the values of ρ_G estimated through both μSXRD observation and hardness measurement will be compared and discussed.

6.3 Experimental

A copper single crystal specimen with a $\langle 111 \rangle$ out-of-plane orientation, in the form of a 2 mm-thick, 10 mm-diameter disk, was purchased from Monocrystals Company. A flat edge was cut along a $\langle 110 \rangle$ direction (normal to a $\langle 112 \rangle$ direction) to provide a reference for the crystal orientation. The indented sample surface was mirror-finished and electropolished. Three-sided Berkovich indentation tests were performed using a Nanoindenter XP^{TM} with the continuous stiffness measurement module. Figure 6.1 shows an optical image of the 5 indentation arrays (each

Fig. 6.1 Optical image of arrays of Berkovich indents on single crystal $\langle 111 \rangle$ Cu with indentation depths ranging from 3 to 0.25 μm

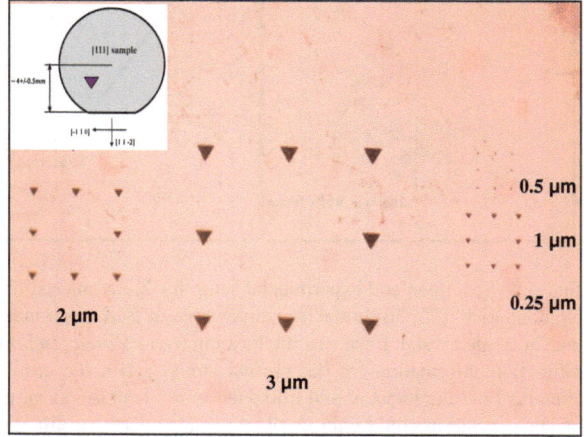

consisting of 8 indents, namely a 3 × 3 array without the center), corresponding to indentation depths of 3, 1.5, 1, 0.5 and 0.25 μm.

The horizontal edge of the Berkovich indents was aimed to be lined up as parallel as possible to the flat edge of the Cu disk, which also means, to the ⟨110⟩ type directions of the Cu single crystal. But due to instrumental limitation, the indent edges of the Berkovich indent were lined up within about 1° of the three ⟨110⟩ type directions in the surface plane of the sample as shown in Fig. 6.2a. For performing the X-ray microdiffraction experiments, the surface of the sample was oriented at an angle of 45° with respect to the incident beam so that the X-ray microbeam could penetrate ∼30 μm below the surface, limited ultimately by X-ray beam attenuation.

Microbeam X-ray diffraction experiments were performed following the methodology described by Yang et al. [35, 36] and used to obtain local lattice rotations associated indentations of different depths. Full X-ray microdiffraction (μXRD) scanning was first conducted covering the overall crystal surface on which the Berkovich indents had been made, as will be described and discussed in the next section. However, only one position on the indented surface was chosen for analysis and discussion. It is as illustrated in Fig. 6.2b. At this position, the X-ray microbeam enters the sample in the middle of the flat face and penetrates the sample for ∼30 μm underneath this flat face; the diffracted intensity diffracts upward into the CCD detector for all positions along the penetration depth. Only at this position does the X-ray beam encounter a relatively simple, single rotation of crystal planes of Cu, such as illustrated in Fig. 6.2c; at other positions, the X-ray beam may

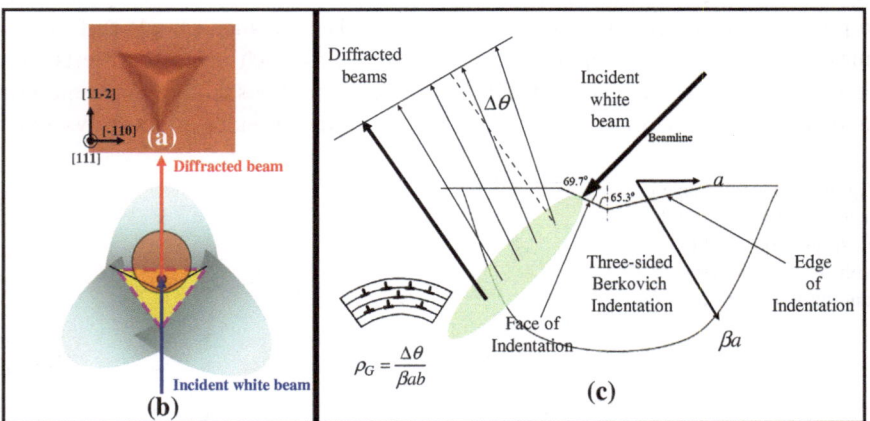

Fig. 6.2 Specimen and experimental setup for X-ray microdiffraction following the methodology by Yang et al. [35, 36]: **a** the horizontal edge of Berkovich indent was aligned with the [−110] of the Cu single crystal; **b** the specific location (*red-colored circle*) which carries the most meaningful diffraction information for the present study, and **c** the cross section of (**b**) on the plane that contains both the incident and diffracted X-ray beams—as an illustration why simple, single axis of lattice rotation could only be observed from such specific position

encounter more than one rotation of crystal planes due to deformations by other faces of the Berkovich indenter.

The white beam X-ray microdiffraction (μXRD) experiment was performed on beamline 12.3.2. at the Advanced Light Source, Berkeley, CA. The sample was mounted on a precision XY Huber stage and oriented at an angle of 45° with respect to the incident beam (Fig. 6.2c). Firstly, the indented sample surface was raster scanned at room temperature under the X-ray beam to provide X-ray microfluorescence (μXRF) mapping, which revealed the Pt markers on the Cu sample to locate the indentation arrays. Then, finer μXRD scanning was conducted on the individual indents using a constant 0.8 μm focused beam size (the full width at half maximum, FWHM = 0.8 μm). As with typical synchrotron experiments, the scanning quantity and quality (resolution) were always balanced against the limited beam time. Only the 3, 1 and 0.25 μm indents were μXRD scanned with step sizes of 2, 1 and 0.5 μm, respectively. For each indentation depth, we scanned 3 individual indentations. The μXRD patterns were collected using a MAR133 X-ray charge-coupled device (CCD) detector and analyzed using the XMAS (X-ray microdiffraction analysis software) software package [32]. For the same experimental setup as shown in Fig. 6.2c, Yang et al. found that [35–37], even after penetrating a copper sample as deep as 30–50 μm, the incident X-ray beam (with an energy range of 8–25 keV) can still generate detectable diffracted beams [38], indicating that the X-ray (with an energy range of 8–25 keV) penetration length for copper is at least 30–50 μm [38]. In our μXRD experiments performed on beamline 12.3.2. at the Advanced Light Source, the incident X-ray beam had an energy range of 5–14 keV, i.e. a lower energy range compared to that in Yang's experiments, and the corresponding maximum X-ray penetration length for copper in our experiments would be around 30 μm.

6.4 Results and Discussion

6.4.1 Mapping of Laue Peak Streaking on Individual Indents

We first describe the μXRD scanning results of the individual indents. The μXRD scanning provides us with the mapping of the (111) Laue diffraction spot from the indented single crystal with the incident X-ray beam located at various positions on the indented surface. Figure 6.3 shows mapping for one of the 3 μm indents. Each of the individual images of the Laue diffraction spots in Fig. 6.3 represents a (diffracted) intensity contour 2-D χ–2θ coordinate system (i.e. the 2-D coordinate in the diffractometer coordinate system).

First and foremost, this mapping shows the difference between the diffraction spots coming from the deformed area of the crystal near the indent versus the undeformed crystal. The outer circle in Fig. 6.3 represents the estimated plastic zone size (βa), whereas the inner circle represents the calculated contact radius, a,

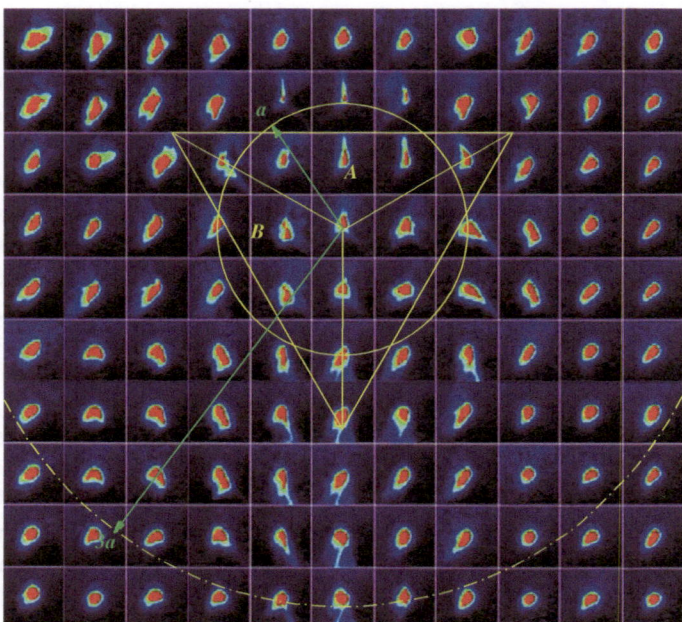

Fig. 6.3 Mapping of the (111) Laue spot of the indented area of the crystal as well as areas surrounding it. Distance between images in this map is 4 μm (even though the scanning was done in 2 μm step size; this was to improve clarity and simplicity for presentation). The *yellow small* and *large circles* represent the contact and the plastic deformation zones, respectively, and the *yellow triangle* represents the Berkovich-indented surface

with an equivalent conical indenter, and the triangle represents the indented surface by the Berkovich indenter (For this 3 μm indent, contact radius, a, is 8.38 μm; β was taken as 3 for annealed Cu [17], the estimated plastic zone radius, βa, was 25 μm; and length of the Berkovich edge is 22.6 μm). For the regions far away from any indents (not shown in Fig. 6.3), the (111) Laue spots are circular without directional streaking, similar to the Laue spot at the lower right corner in Fig. 6.3. On the other hand, Fig. 6.3 shows various types and different extents of streaking in Laue spots, indicating the complexity of indentation-induced deformation.

Furthermore, within the deformed area, the mapping shows complex plastic deformation in the crystal often involving more than one rotation axis. This manifests itself in the form of broadening of the Laue diffraction spots (streaking) in various directions. For example, the crystal volume under the Berkovich tip would have a complex dislocation structure, with fractions of the same volume rotated about at least three different axes. However, as has been suggested by Yang et al. [35, 36], and also briefly discussed above, one particular volume of this indented crystal is expected to show a simple rotation about a single axis. Our observation is consistent with this suggestion, and the mapping in Fig. 6.3 shows only streaking in a single direction in those particular positions on the indented surface, consistent with our illustration in Fig. 6.2b. As this single crystal rotation represents the deformation associated with one face of the Berkovich indentation, the streaked diffracted intensity could be used as a direct measure of strain gradient introduced at a particular depth of indentation. Here, and hereafter, all of our attention will be focused on the streaked diffraction spots in this position for the indents on the indented surface.

6.4.2 Comparison of Laue Peak Streaking for Different Indentation Depths

For each of the three indentation depths analyzed (3, 1 and 0.25 μm), we selected the particular (111) Laue diffraction spots coming from the above-specified position (we had three data sets for each indentation depth), and chose one representative diffraction spot for each depth of indentation. Figure 6.4 shows the streaked Laue diffraction spots for indentation depths of 3, 1 and 0.25 μm in the forms of 3-D images, 2-D χ–θ contour plots, and intensity profiles along the direction of streaking

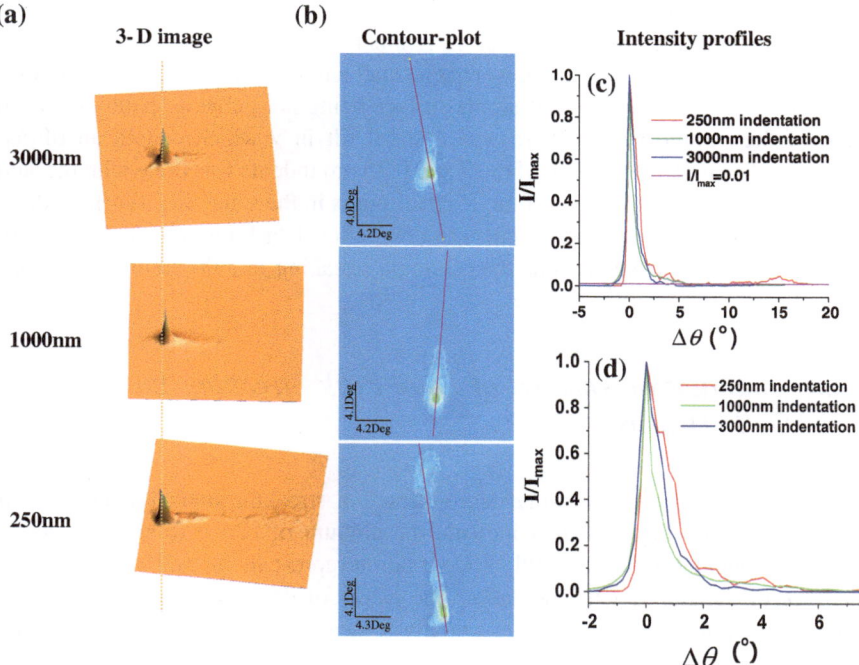

Fig. 6.4 The representative extent of streaking of Laue diffraction peaks from 3, 1, and 0.25 μm indentation depths: **a** 3-D images of normalized intensities versus χ versus θ; **b** 2-D χ–θ contour plot; **c** Profiles of the streaked diffracted intensities (normalized) following the *red lines* in (**b**) which were defined as streaking directions; **d** Profiles of only the main peaks

(red line in the contour plot); Fig. 6.4a–d, respectively. Figure 6.4c shows the overall intensity profile covering the full streaking, whereas Fig. 6.4d focuses only on the main peaks. This intensity profile suggests that the extent of streaking of the Laue diffraction spots for the three different indents are very similar; i.e. they lead to streaking of about $\Delta\theta \sim 5°$. This is useful observation as the extent of streaking gives a measure of density of GNDs, or equivalently the strain gradients. As explained below, these experimental GND density measurements can be correlated with the derived GND densities from the hardness measurements (based on the revised Nix and Gao model [16, 17]).

A heightened diffracted intensity was also observed in Fig. 6.4c at around $\Delta\theta = 15°$, which belongs to the streaking of the Laue diffraction spot of the smallest indent (0.25 μm). This is also shown in Fig. 6.4a, b as scattered intensities on the background; the other two larger indents (3 and 1 μm) show a clear background (Fig. 6.4a, b). We think this is an artifact of the very small plastic zone size, and have excluded it from the main peak, as indicated in Fig. 6.4d. For the 0.25 μm-deep indents, the focused X-ray beam (FWHM = 0.8 μm) samples the entire deformed crystal volume with its complicated, multiple rotational axes giving rise to extra scattering that shows up as heightened intensities in the background (which could not be removed by routine background removal procedures). In the larger indents, the 0.8 μm X-ray beam covered only rather small parts of the entire deformed volume (due to much larger plastic zone sizes) at any given position on the indented surface, thus the scattering is much less and the background is clear.

The directions of the streaking were at a small angle from the vertical direction in Fig. 6.4b. This is caused by the sample surface being not perfectly parallel with the edge of the sample stage. There was a small tilt in χ which causes an off-axis streaking. The μXRD scanning of the 3 and 0.25 μm indents was done with the same sample stage setting (thus the off-angle streaking is in the same direction), while the μXRD scanning of the 1 μm indent was done after detaching and re-attaching the sample to the sample stage (thus the off-angle streaking in a different direction).

6.4.3 Quantitative Analysis of Laue Peak Streaking-Based GND Density

We have already obtained X-ray microdiffraction streaking data (for the 3, 1 and 0.25 μm indents; Fig. 6.4) that indicate the amount of lattice rotation associated with each indentation. As the full lattice rotation represents the full deformation by one face of the Berkovich indentation, the extent of this streaking (broadening of Laue diffraction peaks) gives us essentially a measure of ρ_G associated with each indentation.

The relationship between the extent of streaking, $\Delta\theta$, and the curvature of indented crystal, κ, is obvious from Fig. 6.2c, and as has been generally described in the treatment of other cases of plastic deformation in crystals [33, 34, 41]. The relationship can be approximated as

$$\kappa \approx \frac{\Delta\theta}{\beta a},\qquad(6.1)$$

where $\Delta\theta$ is the extent of streaking observed from µXRD experiment, a is the effective contact radius of indentation, and β is the multiplier factor to a, which gives the estimated plastic zone size.

The Cahn-Nye relationship [42, 43] then gives the relation between the curvature of the indented crystal, κ, and the density of geometrically necessary dislocations (GNDs), ρ_G, associated with that curvature

$$\rho_G = \frac{\kappa}{b}\qquad(6.2)$$

Combining Eqs. 6.1 and 6.2, we find

$$\rho_G = \frac{\kappa}{b} \approx \frac{\Delta\theta}{\beta ab}\qquad(6.3)$$

6.4.4 Hardness Measurement and Revised Nix and Gao's GND Density

Hardness measurements were taken during the indentation of all indents (of all indentation depths), and the results were plotted as H^2 against $1/h$ in Fig. 6.5.

Following the work of Stelmashenko et al. [4] and De Guzman et al. [5], Nix and Gao [8] provided a simple explanation for this depth-dependent hardness, in terms

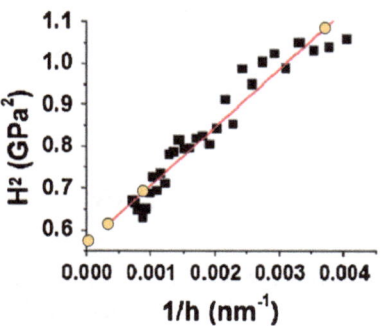

Fig. 6.5 Experimental depth-dependent hardness data; the *orange dots* represent the predicted values of hardness at indentation depths 0.25, 1, 3 µm and bulk samples based on the linear curve fitting of the data using the revised Nix and Gao model [16, 17]

of the geometrically necessary dislocation density as a function of indentation depth. Durst and Göken [16] as well as Feng [17] later modified the model to account primarily for the fact that the plastic zone radius is not equal to the contact radius, as Nix and Gao had assumed. Still the revised model takes the form:

$$\frac{H(h)}{H_0} = \sqrt{1 + \frac{h_0}{h}}$$

(6.4)

which can also be shown equivalently:

$$\frac{H(h)}{H_0} = \sqrt{1 + \frac{\rho_G}{\rho_S}}$$

(6.5)

$$\text{where, } \rho_S = \frac{H_0^2}{3C_H^2 \alpha_t^2 \mu^2 b^2},$$

(6.6)

$H(h)$ is the hardness as a function of h, the depth of the indentation, while H_0 is the limit of the hardness when the indentation depth (h) becomes indefinitely large, and h_0 is a material length scale. In Eq. 6.6, C_H is a constant associated with the plastic zone size, α_t is the Taylor constant, μ is shear strength and b is the magnitude of Burgers vector. As indicated by Eq. 6.4, the plot in Fig. 6.5 shows a linear relationship. The orange dots represent the hardnesses at indentation depths 0.25, 1, 3 μm and the expected hardness at infinite depth (by extrapolation).

Equation 6.5, thus, implies depth-dependent ρ_G (as ρ_S is nominally constant), or in other words depth-dependent strain gradients. The correlation between the hardness numbers and the associated ρ_G had been derived in a complete manner elsewhere [17], and with some rearrangements, the final form such as shown in Eq. 6.7 can be used;

$$\rho_G = \frac{H_0^2}{3C_H^2 \alpha_t^2 \mu^2 b^2} \left\{ \left(\frac{H}{H_0}\right)^2 - 1 \right\}$$

(6.7)

Using this revised Nix and Gao [16, 17] model thus, ρ_G as a function of h, can be derived from the experimental hardness data.

6.4.5 Strain Gradient Plasticity Analysis

We now compute ρ_G versus h, from both the experimental hardness data (using the revised Nix-Gao model) and from the X-ray microdiffraction (streaking) experiments and compare the two results. The comparison is shown in Table 6.1, as well as in Fig. 6.6.

Table 6.1 Comparison of experimental parameters and GND densities representative of the three indents (3, 1, 0.25 μm indentation depths) of interest in the present study

Independent depth (μm)	a (μm)	$\Delta\theta$ (°)	$\rho_G = \frac{\chi}{b} \approx \frac{\Delta\theta}{\beta a b}$ (μm^{-2})	$\rho_G = \frac{H_0^2}{3C_H^2\alpha_s^2\mu^2 b^2}\left\{\left(\frac{H}{H_0}\right)^2 - 1\right\}$ (μm^{-2})
3	8.41	4.6	18.3	22.4
1	2.83	6.5	77.1	67.1
0.25	0.73	5.5	250.3	268.4

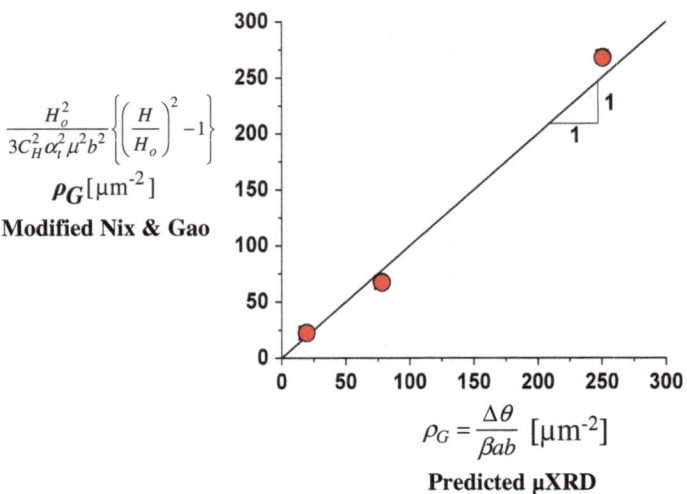

Fig. 6.6 Graphic representation of GND densities obtained through two independent methods and shows self-consistency of the strain gradient plasticity model

The streak length, $\Delta\theta$, was taken at the 1 % threshold ($I/I_{max} = 0.01$) in the normalized intensity plot in Fig. 6.4d. For the calculation of ρ_G from Laue peak streaking, $\beta = 2.35$ was used for the estimated plastic zone size, which is a reasonable value for annealed, electropolished single crystal Cu [17]; corresponding to $\beta = 2.35$, $C_H = 6.6$ was used for the calculation of Eq. 6.3. In addition, the following constants were used: $\alpha_t = 0.4$, $\mu = 44.7$ GPa, $b = 0.2214$ nm for single crystal Cu [17]. The hardness data was fitted using $H_0 = 0.75$ GPa and $h^* = 244$ nm (Fig. 6.5). Also b was taken as the component of the Burgers vector along $\langle 112 \rangle$, which contributes to the strain gradient shown in Fig. 6.7. By applying these conditions, ρ_S, could be calculated using Eq. 6.6 to obtain $\rho_S = 275$ μm^{-2}.

Figure 6.6 shows that the GNDs (and strain gradients) found by these two methods are in close agreement. In fact, given the approximate way that we treat the deformation field under the indenter, the agreement here seems to be better than what we can expect. The figure of merit here is the parameter β for the estimated plastic zone size. If we assume slightly different (but still equally valid) β values, our Cahn-Nye calculation will give slightly different results.

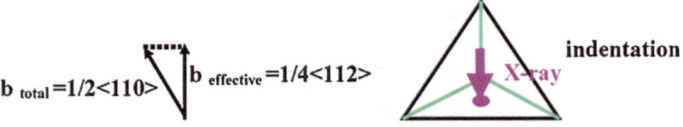

Fig. 6.7 The schematic for the effective burger's vector in the direction of strain gradients with respect to the Berkovich-indented surface

6.5 Conclusions

Using a synchrotron technique involving white-beam X-ray microdiffraction (μXRD), we have observed Laue peak streaking near small indentations in the (111) surface of a copper single crystal. The geometrically necessary dislocation density, ρ_G, computed from the observed streaking increases with decreasing indentation depth, which is in good agreement with ρ_G computed from the observed indentation size effect (ISE) using a revised Nix-Gao model. This finding supports that the ISE is associated with geometrically necessary dislocations and related strain gradients. Moreover, it is demonstrated that μXRD is a good tool for probing the deformation mechanism at the sub-micrometer scale.

References

1. Arzt E (1998) Size effects in materials due to microstructural and dimensional constraints: a comparative review. Acta Mater 46:5611–5626
2. Yu YW, Spaepen F (2003) The yield strength of thin copper films on Kapton. J Appl Phys 95:2991–2997
3. Nix WD (1989) Mechanical properties of thin films. Merall Trans A 20:2217–2245
4. Stelmashenko NA, Walls MG, Brown LM et al (1993) Microindentations on W and Mo oriented single crystals: an STM study. Acta Metall Mater 41:2855–2865
5. De Guzman MS, Neubauer G, Flinn P et al (1993) The role of indentation depth on the measured hardness of materials. Mater Res Soc Proc 308:613
6. Ma Q, Clarke DR (1995) Size dependent hardness of silver single crystals. J Mat Res 10:853–863
7. Poole WJ, Ashby MF, Fleck NA (1996) Micro-hardness of annealed and work-hardened copper polycrystals. Scripta Mat 34:559–564
8. Nix WD, Gao H (1998) Micro-hardness of annealed and work-hardened copper polycrystals. J Mech Phys Solids 46:411–425
9. Gao H, Huang Y (1999) Y.W.D.W.D., Naturwissenschaftler, 86:507
10. Gao H, Huang WD, Nix JW et al (1999) Mechanism-based strain gradient plasticity—I. Theory. J Mech Phys Solids 47:1239–1263
11. Huang Y, Chen JY, Guo TF et al (1999) Analytic and numerical studies on mode I and mode II fracture in elastic-plastic materials with strain gradient effects. Int J Fract 100:1–27
12. Huang Y, Gao H, Nix WD et al (2000) Mechanism-based strain gradient plasticity—II. Analysis. J Mech Phys Solids 48:99–128
13. Huang Y, Xue Z, Gao H et al (2000) A study of microindentation hardness tests by mechanism-based strain gradient plasticity. J Mater Res 15:1786–1796
14. Tymiak NI, Kramer DE, Bahr DF et al (2001) Plastic strain and strain gradients at very small indentation depths. Acta Mater 49:1021–1034
15. Swadener JG, George EP, Pharr GM (2002) The correlation of the indentation size effect measured with indenters of various shapes. J Mech Phys Solids 50:681–694
16. Durst K, Backes B, Goken M (2005) Indentation size effect in metallic materials: correcting for the size of the plastic zone. Scripta Mat 52:1093–1097
17. Feng G (2005) The application of contact mechanics in the study of nanoindentation. Dissertation, Stanford University

18. Durst K, Backes B, Franke O et al (2006) Indentation size effect in metallic materials: modeling strength from pop-in to macroscopic hardness using geometrically necessary dislocations. Acta Mat 54:2547–2555
19. Basinski SJ, Basinski ZS (1979) Plastic deformation and work hardening. In: Nabarro FRN (ed) Dislocations of solids, vol 4: dislocations in metallurgy. North-Holland Publishing Company, Oxford, p 261
20. Castell MR, Howie A, Perovic DD et al (1993) Plastic deformation under microindentations in GaAs/AlAs superlattices. Phil Mag Lett 67:89–93
21. Donovan PE (1989) Plastic flow and fracture of Pd40Ni40P20 metallic glass under an indentor. J Mater Sci 24:523–535
22. Hill R, Lee EH, Tupper SJ (1947) The theory of wedge indentation of ductile materials. Proc R Soc Lond A 188:273–289
23. Mulhearn TO (1959) The deformation of metals by vickers-type pyramidal indenters. J Mech Phys Sol 7:85–88
24. Inkson BJ, Steer T, Mobus G et al (2001) Subsurface nanoindentation deformation of Cu–Al multilayers mapped in 3D by focused ion beam microscopy. J Microscopy 201:256–269
25. Tsui TY, Vlassak J, Nix WD (1999) Indentation plastic displacement field: part I. The case of soft films on hard substrates. J Mater Res 14:2196–2203
26. Tsui TY, Vlassak J, Nix WD (1999) Indentation plastic displacement field: part II. The case of hard films on soft substrates. J Mater Res 14:2204–2209
27. Kiener D, Pippan R, Motz C et al (2006) Microstructural evolution of the deformed volume beneath microindents in tungsten and copper. Acta Mater 54:2801–2811
28. Zaafarani N, Raabe D, Singh RN et al (2006) Three-dimensional investigation of the texture and microstructure below a nanoindent in a Cu single crystal using 3D EBSD and crystal plasticity finite element simulations. Acta Mater 54:1863–1876
29. Viswanathan GB, Lee E, Maher DM et al (2005) Direct observations and analyses of dislocation substructures in the α phase of an α/β Ti-alloy formed by nanoindentation. Acta Mater 53:5101–5115
30. Lloyd SJ, Castellero A, Giuliani F et al (2005) Observations of nanoindents via cross-sectional transmission electron microscopy: a survey of deformation mechanisms. Proc Roayal Soc Math Phys Eng Sci 461:2521–2543
31. Fleck NA, Muller GM, Ashby MF et al (1994) Strain gradient plasticity: theory and experiment. Acta Metall Mat 42:475–487
32. Tamura N, MacDowell AA, Spolenak BC et al (2003) Scanning X-ray microdiffraction with submicrometer white beam for strain/stress and orientation mapping in thin films. J Synchrotron Rad 10:137–143
33. Budiman AS, Tamura N, Valek BC et al (2006) Crystal plasticity in Cu damascene interconnect lines undergoing electromigration as revealed by synchrotron X-Ray microdiffraction. Appl Phys Lett 88:233515
34. Valek BC (2003) X-ray microdiffraction studies of mechanical behavior and electromigration in thin film structures. Dissertation, Stanford University
35. Yang W, Larson BC, Pharr GM et al (2004) Deformation microstructure under microindents in single-crystal Cu using three-dimensional x-ray structural microscopy. J Mater Res 19:66–72
36. Yang W, Larson BC, Pharr M et al (2003) Deformation microstructure under nanoindentations in Cu using 3D X-ray structural microscopy. Mat Res Soc Symp Proc 750:Y8.26
37. Yang W, Larson BC, Pharr M et al (2003) X-ray Microbeam Investigation of Deformation Microstructure in Microindented Cu. Mat Res Soc Symp Proc 779:W5.34
38. Yang W, Larson BC, Tischler JZ et al (2004) Differential-aperture X-ray structural microscopy: a submicron-resolution three-dimensional probe of local microstructure and strain. Micron 35:431–439
39. Barabash R, Ice GE, Larson BC et al (2001) White microbeam diffraction from distorted crystals. Appl Phys Lett 79:749–751
40. Barabash RI, Ice GE, Larson BC et al (2002) Application of white X-ray microbeams for the analysis of dislocation structures. Rev Sci Instr 73:1652–1654

41. Budiman AS, Han SM, Greer JR et al (2007) A search for evidence of strain gradient hardening in Au submicron pillars under uniaxial compression using synchrotron X-ray microdiffraction. Acta Mat 56:602–608
42. Cahn RW (1949) Recrystallization of single crystals after plastic bending. J Inst Met 86:121
43. Nye JF (1953) Some geometrical relations in dislocated crystals. Acta Metall 1:153–162

Chapter 7
Smaller is Stronger: Size Effects in Uniaxially Compressed Au Submicron Single Crystal Pillars

Abstract A study of submicron single crystal Au pillar, before and after uniaxial plastic deformation, is discussed in this chapter. There is no evidence of measurable lattice rotation or lattice bending/polygonization due to the deformation up to a plastic strain of about 35 % and a flow stress of close to 300 MPa. These observations, coupled with other examinations using electron microscopy, suggest that plasticity here is not controlled by strain gradients, but rather by dislocation source starvation.

Keywords Smaller is stronger · Size effect · Micropillars · μXRD · Dislocation starvation

7.1 Introduction

When crystalline materials are mechanically deformed in small volumes, higher stresses are needed for plastic flow. This has been called the "Smaller is Stronger" phenomenon and has been widely observed. Various size-dependent strengthening mechanisms have been proposed to account for such effects, often involving strain gradients. Here we report on a search for strain gradients as a possible source of strength for single-crystal submicron pillars of gold subjected to uniform compression, using a submicron white-beam (Laue) X-ray diffraction technique. We have found both before and after uniaxial compression, no evidence of either significant lattice curvature or sub-grain structure. This is true even after 35 % strain and a high flow stress of 300 MPa were achieved during deformation. These observations suggest that plasticity here is not controlled by strain gradients or substructure hardening, but rather by dislocation source starvation, wherein smaller volumes are stronger because fewer sources of dislocations are available.

© The Author(s) 2015
A.S. Budiman, *Probing Crystal Plasticity at the Nanoscales*,
SpringerBriefs in Applied Sciences and Technology,
DOI 10.1007/978-981-287-335-4_7

103

7.2 Background

Unlike the size effect that we observed and studied in Chap. 6 (indentation size effect), a different kind of intrinsic size effect appears to have also been observed [1] when single crystalline materials in the form of micro pillars are deformed homogenously, without strain gradients. Recently Uchic et al. [1] , Greer et al. [2] and 3. Greer and Nix [3] have shown that micro pillars of various metals with diameters in the micron range, subjected to uniaxial compression, are much stronger than bulk materials. For example, micro pillars of gold ranging in diameter between 200 nm and several microns have been found to have compressive flow strengths as high as 800 MPa, a value ~ 50 times higher than the strength of bulk gold [2, 3]. This suggests that in spite of much progress on size effects on strength there is still no unified theory for plastic deformation at the sub-micron scale. The accounts of strain gradient plasticity, as illustrated in the previous chapter (Chap. 6, appear to break down for the case of micro pillar compression because the geometry of the micro pillar compression is not expected to include externally-imposed plastic strain gradients that might lead to extra hardening for small samples.

While significant macroscopic strain gradients are not expected to develop during the uniform compression of micron sized pillars, we cannot preclude microscopic strain gradients from forming. Nevertheless, even the presence or the absence of macroscopic strain gradients has not been directly and experimentally observed, especially in the case of metallic pillars. Since strain gradients and GNDs are directly related to the local lattice curvature, the technique of Scanning X-ray Microdiffraction (μSXRD) using a focused polychromatic/white synchrotron X-ray beam can be used to determine the density of GNDs. This has proven to be useful in the study of the early stages of electromigration failure in interconnect lines, wherein lattice bending and GNDs are created by electromigration processes [4, 5]. This capability is related to the continuous range of wavelengths in a white X-ray beam, allowing Bragg's Law to be satisfied even when the lattice is locally rotated or bent, resulting in the observation of streaked Laue spots.

Using this approach, we can monitor the change in the Laue diffraction peaks before and after uniaxial compression of a sub-micron single crystal Au pillar. A quantitative analysis of the Laue peak widths then allows us to estimate the density of GNDs in the pillar. The absolute number of geometrically necessary dislocations in the crystal can then be determined using the dimensions of the pillar. A comparison of the numbers of geometrically necessary dislocations before and after the uniaxial compression provides information about the change in microstructure associated with plastic deformation.

The technique of synchrotron-based white beam X-ray diffraction is one of the few methods for detecting and measuring the densities of GNDs in crystalline materials after deformation. Other viable techniques include TEM and Resonant Ultrasound Spectroscopy (RUS) [6].

The synchrotron technique of scanning white beam X-ray microdiffraction has been described thoroughly elsewhere [7]. The power of this technique to study local plasticity and mechanical behavior of materials at small scales stems from the high brilliance of the synchrotron source, as well as the recent advances in X-ray focusing optics (allowing sample mapping at the sub-micron level).

7.3 Experimental

7.3.1 Thin Film of Au on Single Crystal Cr Substrate

The sample consists of a 4-crystal film of gold oriented ⟨111⟩ out-of-plane (Fig. 7.1), and deposited onto a ⟨001⟩ chromium single crystal substrate, in the form of a 2 mm-thick, 10 mm-diameter disk. The native oxide on the surface of the Cr substrate was first removed by ion cleaning in a high vacuum, using the appearance of a RHEED pattern of single crystal Cr to indicate the removal of the surface oxide. A thin layer of gold was then epitaxially deposited onto the bare Cr surface by vapor deposition in the same high vacuum chamber, before moving the sample to a sputtering system to continue growth to a thickness of 1.9 μm. While a ⟨001⟩

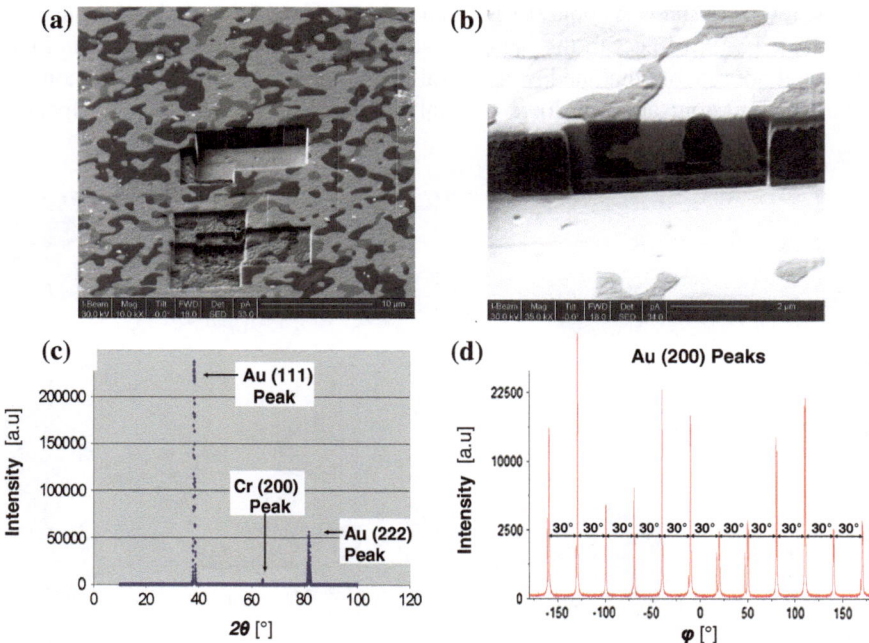

Fig. 7.1 The Au (111) film on Cr (200) substrate: **a** 4 variants of Au ⟨111⟩ out-of-plane crystals were observed in the planar view, as well as **b** through thickness of film (using ion beam contrast imaging inside an SEM at a 52° tilted angle); **c** Symmetric X-ray diffraction gave the Au (111) peaks of the film and the Cr (200) peak of the single crystal Cr substrate; however **d** non-symmetric X-ray diffraction (by tilting the sample 54.74° to get Au (200) peaks, and then scanning φ for 360°) gave 4 sets of three Au (200) peaks each separated by 120° angle, indicating the 4-crystal Au <111> film

orientation was expected for the gold film on the basis of interface energy considerations, the low energy of the (111) surface must have caused the $\langle 111 \rangle$ orientation to be selected. For this orientation, a nearly perfect lattice match is achieved along a $\langle 110 \rangle$ direction in the (111) surface of Au and a $\langle 100 \rangle$ direction in the (001) surface of Cr.

7.3.2 Fabrication and Uniaxial Compression of Submicron Au Pillar

The pillars were fabricated utilizing the method of FIB machining, following the approach developed by Greer et al. [3]. Circular craters 30 μm in diameter were first carved out of the gold film, leaving behind only the sub-micron pillars at the centers of the craters (Figs. 7.2a, b). Using scanning electron microscopy, we determined that the Au pillar (Fig. 7.2c) was a bicrystal of Au with top and bottom crystals having different in-plane orientations. The top crystal was about 1.1 μm in height and became the focus of our experiment. It has diameter of 0.58 μm leading to a height to diameter ratio of close to 2:1 for that crystal.

The lower crystal was about 0.8 μm in height (and also 0.58 μm in diameter) and misoriented with respect to the upper crystal by about 30° (in-plane orientations). We could not rule out that the lower crystal might have deformed too in response to pillar compression; also the lower crystal might present a barrier to dislocation

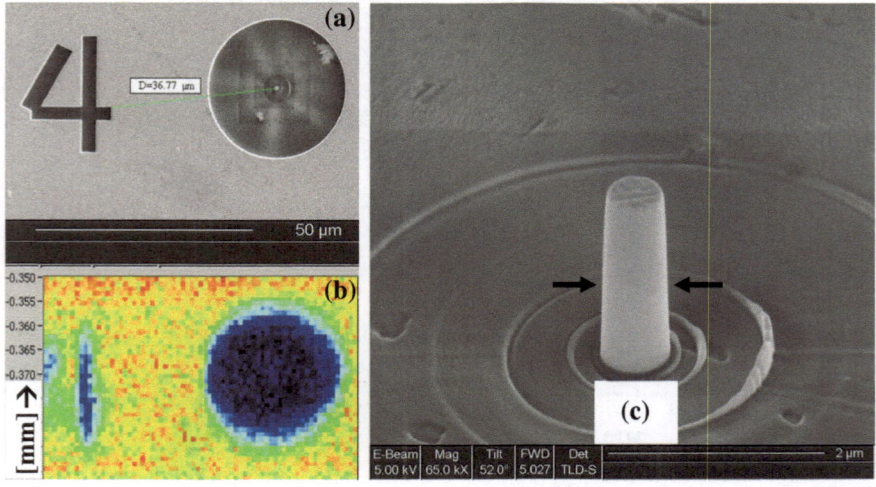

Fig. 7.2 The submicron Au pillar specimen; **a** the crater in the middle of which stands the pillar, with the identifying mark (number "4") on the *left* SEM image, and **b** synchrotron white-beam X-ray microFluorescence (XRF) scan; **c** a $\langle 111 \rangle$-oriented gold pillar machined in the FIB (pillar diameter = 580 nm, pillar height = 1.9 μm (total) and ∼1.1 μm (*the upper crystal*); a slight color contrast signifying differently oriented crystals between the *upper* and *bottom crystals* is visible upon careful inspection, as indicated by the *arrows* pointing to the interface)

motion in the upper crystal. However, this barrier would not be stronger than having the Cr substrate on the bottom of the upper crystal.

The uniaxial compression testing of these submicron pillars was conducted using an MTS Nanoindenter XP with a flat punch diamond tip, following the methodology described by Han [8]. The nanoindenter, which is a load-controlled instrument, was programmed to perform a nominally displacement-controlled test. In this method, the displacement rate is calculated continuously during the compression test, based on the measured displacement and time. When the measured displacement rate is below a specified value, the load is adjusted to maintain that particular displacement rate. This method is designed to simulate a constant displacement rate. Load-displacement data were collected in the continuous stiffness measurement (CSM) mode of the instrument. The data obtained during compression were then converted to uniaxial stresses and strains using the assumption that the plastic volume is conserved throughout this mostly-homogeneous deformation.

7.3.3 White-Beam X-ray Microdiffraction Experiment

The white beam X-ray microdiffraction experiment (Fig. 7.3) was performed on beamline 12.3.2. at the Advanced Light Source, Berkeley, CA. The sample was mounted on a precision XY Huber stage and the pillar of interest was raster scanned at room temperature under the X-ray beam before and after the uniaxial compression (ex situ); this provided X-ray microFluorescence (μXRF) and X-ray

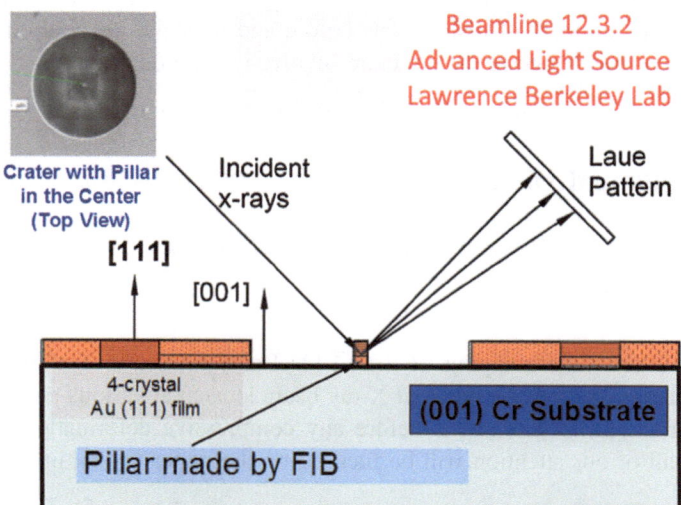

Fig. 7.3 A schematic illustration of the synchrotron white-beam X-ray microdiffraction experiments (conducted at the ALS beamline 12.3.2, Lawrence Berkeley National Lab) on the Au submicron pillar on Cr substrate

microDiffraction (μXRD) scans for the area near the pillar. The μXRD patterns were collected using a MAR133 X-ray CCD detector and analyzed using the XMAS software package.

The μXRF scan was conducted first for precise positioning of the Au pillar prior to the μXRD scan. The μXRF scan was made using 1 μm step sizes to cover a large area of typically 70 × 70 μm (to include not only the 30 μm diameter crater, but also the identifying mark—number "4" in Fig. 7.1). The circular crater was used, first, to clearly locate the position of the pillar (as the pillar was fabricated at the center of the crater), and, secondly, to partially separate the diffraction signal of the pillar from that of the surrounding gold film, such that only those diffracted beams from the pillar (and not those of the surrounding film) can be studied.

Once the Au pillar was located and identified, an μXRD scan was conducted to obtain diffraction data primarily from the Au crystal/pillar. The typical scan for this purpose was made with 1 μm step sizes, 50 steps across the diameter of the crater (50 μm scan length) and 10 steps along its orthogonal direction (10 μm scan length), making it a wide band of 50 × 10 μm. This scan area was designed to include not only the pillar, but also the boundaries of the crater with the surrounding Au film as positional references. This μXRD scan involved the collection of 500 CCD frames. A complete set of CCD frames took about 4–5 h to collect. The exposure time was 5 s, in addition to about 10 s of electronic readout time for each frame.

With μXRD, we monitored the change in the Laue diffraction images before and after uniaxial compression of a single crystal submicron Au pillar. A quantitative analysis of the Laue peak widths allows us to estimate the density of GNDs in the submicron single crystal pillars. The exact geometries of the pillars being known, the absolute number of dislocations in the single crystal can be derived. A comparison of the numbers of dislocations before and after the uniaxial compression would unveil the change in the structure involved in the deformation.

7.4 Results and Discussion

7.4.1 Diffraction Intensity Mapping: Pillar Location Identification

Figure 7.4 shows the mapping of the ($\bar{3}$ 11) Laue diffraction spot from the top crystal of the pillar with the incident X-ray beam located at various positions in the vicinity of the pillar and crater, before any compressive deformation. Here, and hereafter, all of our attention will be focused on the upper crystal in the Au pillar structure.

Each of the individual images of the Laue diffraction spots in Fig. 7.4 represents a (diffracted) intensity contour in a 2-D χ-2θ coordinate system (i.e., the 2-D coordinate in the diffractometer coordinate system). As we set the threshold of the

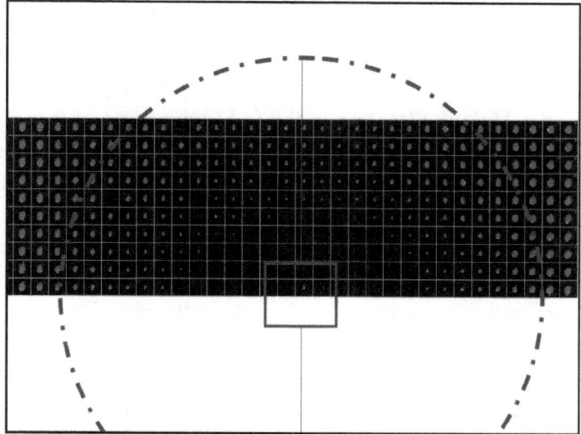

Fig. 7.4 A mapping of the ($\overline{3}$11) Laue spot of the upper crystal of the pillar in the areas surrounding the pillar and the crater; step size = 1 μm. The *dashed* and *dotted* (*blue*) *circle* represents the 30 μm-crater, and as we expect, exactly in the middle of it, stands the pillar as marked by the *rectangular solid line box*

lower-bound intensity display to be the same for all images, the difference in peak size/width in the mapping indicates a difference in the absolute diffracted intensity of the Au crystal volume at a particular position in the map. Thus, the size of the red "dots" is directly related to the diffracted volume of Au crystal. The bigger red "dots" on the left-hand and right-hand sides of this map clearly represent the surrounding Au films, while the smaller red dots in the middle area represent the crater (close to zero diffracting volume). Obviously, as there is no Au crystal away from the center of the crater, there ought to be absolutely zero diffraction intensity in this area. However, because of the Lorentzian shape of the incoming focused X-ray beam, the tails of the beam extend beyond (up to tens of microns) the nominal FWHM (0.8 μm) of the beam. Therefore, even though the beam is focused on a particular location in the crater, there is still a very small fraction of diffraction intensity coming from the surrounding film picked up by the tails of the X-ray beam.

The diffraction map in Fig. 7.4 also indicates unambiguously the exact location of the pillar itself (marked by the solid line box in Fig. 7.4). After subtracting out the diffraction intensity by the tails of the X-ray beam, a significant diffraction intensity was still left on the location at the centre of the solid line box (i.e., the pillar). This is so as there is indeed slightly more volume of Au crystal to diffract at the location of the pillar, but this intensity is not as high as in the surrounding film areas (thus the relative size of the peak). The location associated with the slightly bigger Laue diffraction spot also coincides with the center of the crater as identified by dashed and dotted circle in the map of Fig. 7.4

7.4.2 Stress-Strain Behavior of Pillar Uniaxial Compression

Figure 7.5a shows the stress-strain curves of the 0.58 μm Au pillar obtained during the compression testing. Uniaxial loading in the $\langle 111 \rangle$ direction of the Au crystal/pillar, corresponding to a high-symmetry orientation, would result in the activation of multiple slip systems, with the pillar deforming uniformly around its diameter as it is compressed. The flow stress reaches value as high as 280 MPa. This is close to 10 times the yield stress of gold in bulk, and falls consistently in the flow stress versus pillar diameter chart described by Greer and Nix [3]. In this $\langle 111 \rangle$ loading orientation, and despite the presence of the end constraints, the pillar remains centrally-loaded and preserves its cylindrical shape throughout the deformation process as shown in Fig. 7.5b. While upon further inspection we could not observe slip steps on the surface of the deformed pillar, we do however observe a visible slip marking (not shown very clearly in this particular SEM image in Fig. 7.5b) that appears to be consistent with the trace of a {111} plane of the pillar crystal. The angle made by the plane causing the trace was within a few degrees of the expected trace of a {111} plane, after correcting for subsequent plastic deformation of the pillar. The final diameter of the pillar after the uniaxial compression is 0.67 μm, which represents a total strain of close to 35 %.

7.4.3 Laue Diffraction Peak Shapes: Undeformed Versus Deformed

A diffraction scan was again taken covering the deformed pillar and the surrounding area (including the crater border with the surrounding Au film) similarly to the

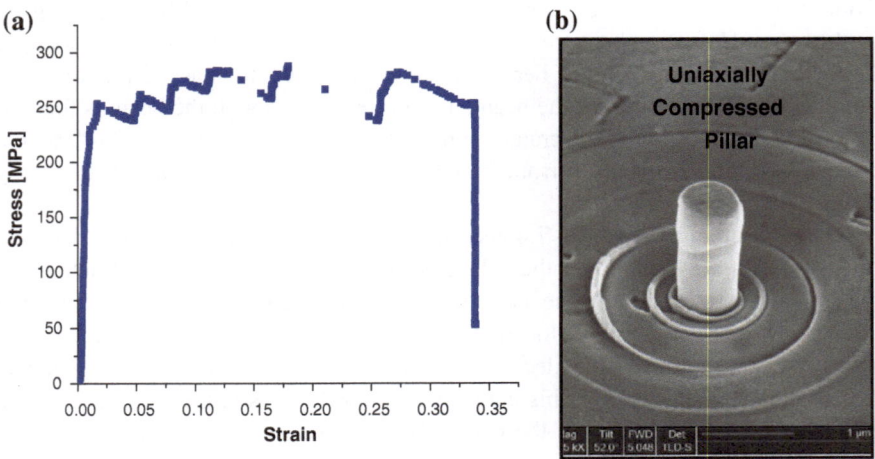

Fig. 7.5 Stress-strain behavior of $\langle 111 \rangle$ -oriented Au single crystal submicron pillar: **a** flow stress increases significantly beyond its typical bulk values; **b** SEM image of a uniaxially compressed pillar after deformation

diffraction map in Fig. 7.4. Following the methodology described above, we again identify the location of the deformed pillar, and subsequently we select a particular Laue diffraction spot (in the case shown here, ($\bar{3}$ 11) diffraction spot) associated with the location of the deformed pillar for further quantitative analysis and comparison. We subtracted the background intensity and checked that there is no Au crystal rotation involved (which would have manifested in the shift of the position of the Au Laue peak with respect to the Laue diffraction pattern of the chromium substrate reference), in order to be able to directly compare the ($\bar{3}$ 11) Laue spots before and after deformation and infer what happened to the pillar crystal during the deformation process. Figure 7.6 shows the data of the pillar crystal (SEM images, Laue diffraction spots, and Intensity profiles) for both undeformed and deformed states side-by-side.

Figure 7.6b shows that both ($\bar{3}$ 11) Laue spots have the same shape and that they are both rounded—not broadened toward a certain direction (streaked). This rounded shape is typical of an undeformed crystal, whereas broadening of a Laue diffraction spot in a certain direction (streaking) would have been associated with the presence of strain gradients in the deformation volume. A more detailed treatment of the streak length and its correlation with the curved, or in general,

Fig. 7.6 Side-by-side comparison between undeformed and deformed states of the pillar crystal; **a** SEM image of the pillar, **b** Laue diffraction spot ($\bar{3}$ 11), and **c** quantitative analysis

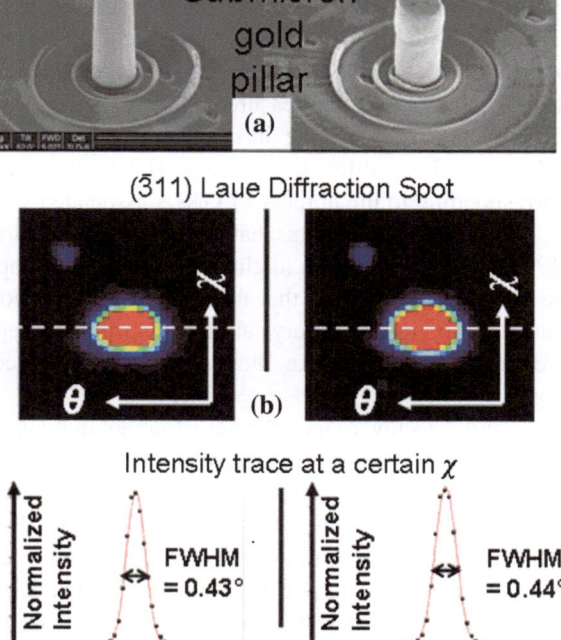

plastically deformed crystal has been described elsewhere [4, 5]. In addition, streaked Laue diffraction spots have also been observed recently in as-fabricated silicon micro pillars [9].

In Fig. 7.6c we take the intensity traces along a particular χ to study the Laue diffraction peak profile more quantitatively (here the χ angle is simply the angle orthogonal to the 2θ angle). The profiles were fitted with Lorentzian curves. The measured FWHMs (full width half maximum) of both profiles show that there is an increase of 0.01° in the angular width. However, this difference is still within the experimental error bar of the instrumentation [7, 10] rendering the two measurements statistically identical. The angular resolution of this technique was calculated using a few assumptions on the experimental sample setup with respect to the CCD camera and on the capability of the indexing code [10], which are applicable to our micro pillar compression experiments. The technique is sensitive to local lattice rotation, and thus this angular detection limit is applicable to geometrically necessary dislocations (GND).

7.4.4 Limitation of the Technique: Quantitative Analysis of GND Density

In terms of the capability of this technique to detect GNDs, it is also limited by the instrumental broadening inherent in the observed Laue diffractions spots. One practical way to estimate the extent of instrumental broadening in our experiments is to take FWHM measurements of Laue diffraction spots coming from a silicon substrate/wafer, as such single crystals are very close to being defect-free and 100 % pure. Based on measurements on such silicon wafer substrates, for similar experimental settings, the instrumental broadening contribution to the observed FWHM of Laue diffraction spots is 0.06°. This places a limit on our technique corresponding to the number of GNDs associated with a peak broadening of 0.06°.

This limitation indicates that relative lattice rotations smaller than $\Delta\theta = 0.06°$, or $\Delta\theta = 10^{-3}$ radians, which might be produced by compressive deformation could not be detected. Converting that measurement to the possible number of dislocations that could be left in the crystal after deformation depends on how the dislocations are distributed. To make these calculations we consider the model shown in Fig. 7.7.

Figure 7.7 shows a model where different domains of size Δs are each assumed to be occupied by like-signed edge dislocations, leading to a local lattice curvature of magnitude $|\kappa|$. Thus within each domain there are only geometrically necessary dislocations. Taken over the whole crystal the dislocations can be regarded as statistically stored dislocations. We know from the Cahn-Nye relationship [11, 12] that the local geometrically necessary dislocation density is $\rho = |\kappa|/b$ where b is the magnitude of the Burgers vector. But the local curvature is $|\kappa| = d\theta/ds \approx \Delta\theta/\Delta s$, so that $\rho = \Delta\theta/b\Delta s$. With this model the total number of dislocations in the entire

Fig. 7.7 A model of the pillar single crystal consisting of different domains of size Δs, each assumed to be occupied by like-signed edge dislocations, leading to a local lattice curvature of magnitude $|\kappa|$

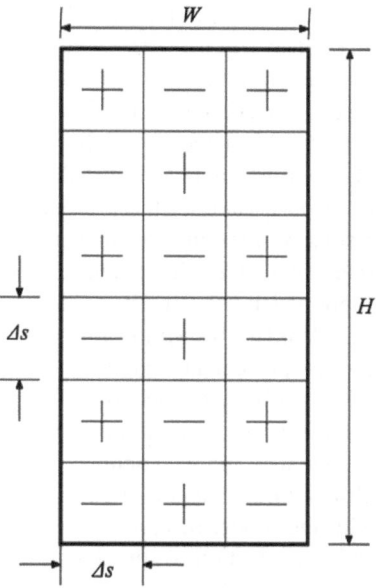

crystal would be $n = \rho W H = W H \Delta \theta / b \Delta s$. Taking the dimensions of the upper crystal to be $W = 600\,\text{nm}$ and $H = 1100\,\text{nm}$, and the Burgers vector to be $b = 0.3\,\text{nm}$ and using the X-ray broadening resolution of $\Delta \theta = 10^{-3}$ we find $n = 2.2 \text{x} 10^3 / \Delta s$. When the domain size is the same as the width of the crystal, $\Delta s = 600\,\text{nm}$, this leads to 3–4 dislocations left in the crystal after deformation. The expected number of dislocations that could be left in the crystal after deformation naturally increases with decreasing domain size. In the limit, a very high density of statistically stored dislocations cannot be ruled out by these experiments, though escape of these dislocations from the nearby free surfaces makes this unlikely.

7.4.5 Dislocation Starvation and Dislocation Nucleation-Controlled Plasticity

Thus the fact that the Laue diffraction spot in the deformed pillar is not streaked suggests that there might not be significant macroscopic strain gradients created during the uniaxial compression of the pillar (only 3–4 GNDs left in the crystal after deformation when the domain size is taken as the same as the width of the crystal, $\Delta s = 600\,\text{nm}$). However, within the resolution and limitation of our technique, we could not preclude the existence of microscopic strain gradients within the inspected volume, as illustrated in Fig. 7.7. In addition, as we find neither a shift in the absolute position of the Laue spots, nor a split, we can further infer that there is no crystal rotation or polygonization (formation of subgrain structures) in the pillar crystal upon the deformation. This is an important observation considering the huge amount of strain (~ 35 %) to which the crystal was subjected.

This observation is consistent with the earlier TEM observations on a deformed gold pillar conducted by Greer and Nix [3]. Their TEM results showed that there were only 2 dislocations left in their pillar after deformation and that they were both of a non-movable type for the uniaxial compressive loading of their experiment. Both observations (the present Laue X-ray microdiffraction and the earlier TEM results) support the idea that sub-micron single crystal gold pillars are nearly defect-free even after significant plastic deformation. This view was recently further supported by the in situ TEM observations of the compression of single crystal nickel pillars by Minor's group at Berkeley Lab [13]. They found that the dislocations present in undeformed pillars (including some dislocation loops near the pillar surfaces created by FIB damage) quickly escaped from the pillar during compressive deformation, leaving the pillar free of dislocations after compression.

We may now conclude that the present white-beam X-ray microdiffraction observations, supported by the closely related TEM results [3, 13], are consistent with the model of hardening of small crystals by dislocation starvation and dislocation nucleation or source-controlled plasticity, as suggested by Greer and Nix [3]. In ordinary plasticity (i.e., in typical, bulk samples), dislocation motion leads to dislocation multiplication by various cross-slip processes, invariably leading to softening before strain hardening occurs through elastic interaction of dislocations. However in small samples, such as the sub-micron Au single crystal pillar under study here, dislocations can travel only very small distances before annihilating at free surfaces, thereby reducing the overall dislocation multiplication rate. The central idea is that, as dislocations leave the crystal more frequently than they multiply, the crystal can quickly reach a dislocation-starved state. When such a state is reached, continued loading would force other, harder sources of dislocations to be activated in the crystal, leading to the abrupt rise in the measured flow stress (i.e., hardening).

7.5 Conclusions

Using synchrotron white-beam X-ray sub-micron diffraction, we have studied a submicron single crystal Au pillar, before and after uniaxial plastic deformation, and found no evidence of measurable lattice rotation or lattice curvature caused by the deformation, even though a plastic strain of about 35 % was imposed and a high flow stress of close to 300 MPa was achieved in the course of deformation. These observations, coupled with other examinations using electron microscopy, suggest that plasticity here is not controlled by strain gradients, but rather by dislocation source starvation, with smaller volumes being stronger because fewer sources of dislocations are available. The central idea of this model is that for very small crystals, dislocations leave the crystal more frequently than they multiply, forcing other, harder sources of dislocations to be activated. Understanding and controlling the mechanical properties of materials on this scale may thus lead to new and more robust nanomechanical structures and devices.

References

1. Uchic MD, Dimiduk DM, Florando JN et al (2004) Sample dimensions influence strength and crystal plasticity. Science 305:986–989
2. Greer JR, Oliver WC, Nix WD (2005) Size dependence of mechanical properties of gold at the micron scale in the absence of strain gradients. Acta Mater 53:1821–1830
3. Greer JR, Nix WD (2006) Nanoscale gold pillars strengthened through dislocation starvation. Phys Rev B 73:245410
4. Valek BC (2003) X-ray microdiffraction studies of mechanical behavior and electromigration in thin film structures. Dissertation, Stanford University
5. Budiman AS, Tamura N, Valek BC et al (2006) Crystal plasticity in Cu damascene interconnect lines undergoing electromigration as revealed by synchrotron X-ray microdiffraction. Appl Phys Lett 88:233515
6. Maurel A, Mercier J, Lund F (2004) Elastic wave propagation through a random array of dislocations. Phys Rev B 70:024303
7. Tamura N, MacDowell AA, Spolenak BC et al (2003) Scanning X-ray microdiffraction with submicrometer white beam for strain/stress and orientation mapping in thin films. J Synchrotron Rad 10:137–143
8. S. M. Han (2006) Methodologies in determining mechanical properties of thin films using nanoindentation. Dissertation, Stanford University
9. Maaβ R, Grolimund D, Petegem S et al (2006) Defect structure in micropillars using X-ray microdiffraction. Appl Phys Lett 89:151905
10. MacDowell AA, Celestre RS, Tamura N et al (2001) Submicron X-ray diffraction. Nucl Inst Meth Phys Res 467–468:936–943
11. Cahn RW (1949) Recrystallization of single crystals after plastic bending. J Inst Met 86:121
12. Nye JF (1953) Some geometrical relations in dislocated crystals. Acta Metall 1:153–162
13. Shan ZW, Mishra R, Asif SAS et al (2007) Mechanical annealing and source-limited deformation in submicrometre-diameter Ni crystals. Nature Mat 7:115–119

Chapter 8
Conclusions

Abstract Our present understanding of the electromigration induced plasticity, indentation size effect and small scale plasticity at the uniaxial compressive stress are summarized in this chapter.

In the electromigration studies using real industry-relevant copper interconnect test structures, we have unraveled a new phenomenon which has not so far been taken into consideration and which might thus change our current understanding of the electromigration degradation mechanisms. This unraveling of these plastic behaviors of copper polycrystalline lines undergoing high current density flux was made possible by the white-beam nature of the X-ray source used in the μSXRD technique. The current understanding of the electromigration phenomenon so far has only included the elastic response of the metallic grains against the global atomic migration in the interconnects. Our results show that this might not be the whole story as plasticity comes into the picture. Furthermore, the plasticity that we observe is of a particular direction corresponding to the direction transverse to the electron flow direction in the line. At some extent of plasticity, this particular configuration could lead to enough additional electromigration flux to start changing the kinetic of the electromigration lifetime prediction. This would certainly have important industrial as well as fundamental implications.

Indentation Size Effect (ISE) has been explained in terms of strain gradients. Smaller indentation leads to stronger lattice rotation or curvature as the same amount of rotation has to be accommodated within the smaller volume of deformation. This would in turn give the higher density of geometrically necessary dislocations (GND's) in the smaller indentation, and thus explains the small scale hardening. Using the white-beam X-ray microdiffraction, which is sensitive of local lattice rotation, it was shown in this book how to directly observe and measure this dependence, and compare it with theoretical values. The results show a reasonable agreement, and thus support that the ISE is associated with GND's and related strain gradients, and demonstrate the unique capability of the μSXRD technique as a local plasticity probe in the submicron and nanometer scales.

While the size-dependence of the hardness of metals has been described in terms of the geometrically necessary dislocations (GND) created in the crystal, such as in

© The Author(s) 2015
A.S. Budiman, *Probing Crystal Plasticity at the Nanoscales*,
SpringerBriefs in Applied Sciences and Technology,
DOI 10.1007/978-981-287-335-4_8

the indentation above, such accounts break down when the size of the deformation volume begins to approach the spacing of individual dislocations. In this domain, the nucleation of dislocations appears to be more important than strain gradients. In an effort to shed light on these topics, uniaxial compression experiments on single-crystal submicron Au pillars made by focused ion-beam machining were conducted. These experiments involve small deformation volumes and strong size effects, yet no evidence of significant lattice curvature is observed. These observations, coupled with other examinations using electron microscopy, suggest that plasticity here is not controlled by strain gradients, but rather by dislocation source starvation, with smaller volumes being stronger because fewer sources of dislocations are available. These results again underline the unique capability of the μSXRD technique as a detector of the presence or rather, in this case, the absence of strain gradients. The technique has proved useful in the understanding and controlling the mechanical properties of materials on small scales, which could well lead to new and more robust nanomechanical structures and devices.